# The Territorial Organization of Variety

## Cooperation and Competition in Bordeaux, Napa and Chianti Classico

JERRY PATCHELL
*Hong Kong University of Science and Technology, Hong Kong*

Routledge
Taylor & Francis Group

LONDON AND NEW YORK

First published 2011 by Ashgate Publishing

Published 2016 by Routledge
2 Park Square, Milton Park, Abingdon, Oxon OX14 4RN
711 Third Avenue, New York, NY 10017, USA

*Routledge is an imprint of the Taylor & Francis Group, an informa business*

**British Library Cataloguing in Publication Data**
Patchell, Jerry.
  The territorial organization of variety : cooperation and
  competition in Bordeaux, Napa and Chianti Classico.
  1. Wine districts. 2. Wine and wine making--France--
  Bordeaux. 3. Wine and wine making--California--Napa
  Valley. 4. Wine and wine making--Italy--Chianti.
  5. Cooperative wineries. 6. Cooperative marketing of farm
  products. 7. Branding (Marketing)
  I. Title
  338.4'76632-dc22

**Library of Congress Cataloging-in-Publication Data**
Patchell, Jerry.
  The territorial organization of variety : cooperation and competition in Bordeaux, Napa
and Chianti Classico / by Jerry Patchell.
     p. cm.
  Includes index.
  ISBN 978-1-4094-1145-1 (hardback)
  1. Wine industry. 2. Vintners. I. Title.
  HD9370.5.P38 2010
  338.4'76632--dc22

                                                                    2010038067

ISBN 9781409411451 (hbk)

# Contents

# The Territorial Organization of Variety
## Cooperation and Competition in Bordeaux, Napa and Chianti Classico

JERRY PATCHELL
*Hong Kong University of Science and Technology, Hong Kong*

LONDON AND NEW YORK

First published 2011 by Ashgate Publishing

Published 2016 by Routledge
2 Park Square, Milton Park, Abingdon, Oxon OX14 4RN
711 Third Avenue, New York, NY 10017, USA

*Routledge is an imprint of the Taylor & Francis Group, an informa business*

**British Library Cataloguing in Publication Data**
Patchell, Jerry.
  The territorial organization of variety : cooperation and
  competition in Bordeaux, Napa and Chianti Classico.
  1. Wine districts. 2. Wine and wine making--France--
  Bordeaux. 3. Wine and wine making--California--Napa
  Valley. 4. Wine and wine making--Italy--Chianti.
  5. Cooperative wineries. 6. Cooperative marketing of farm
  products. 7. Branding (Marketing)
  I. Title
  338.4'76632-dc22

**Library of Congress Cataloging-in-Publication Data**
Patchell, Jerry.
  The territorial organization of variety : cooperation and competition in Bordeaux, Napa
and Chianti Classico / by Jerry Patchell.
       p. cm.
  Includes index.
  ISBN 978-1-4094-1145-1 (hardback)
  1. Wine industry. 2. Vintners. I. Title.
  HD9370.5.P38 2010
  338.4'76632--dc22

                                                                          2010038067

ISBN 9781409411451 (hbk)

# Contents

# List of Figures and Tables

**List of Figures**

**List of Tables**

# Preface

The wine industry is unique. No other industry produces such a great variety of products, in so remarkable a manner, and with such positive social and environmental influences. Not only are tens of thousands of different brands produced, but the great majority of these are created by small entrepreneurs called winegrowers. They are farmers who process and market their crops. They have cracked the most intractable dilemma in agriculture, generating a high level of value added from a small plot of land. The value comes from wines that reflect their own creativity, the traditions of their communities, and the distinct characteristics of their land and microclimates.

Whether driven by a passion to express their own creativity or the desire to add value to their products, the efforts of winegrowers invite us to explore a wealth of variety. Wine introduces vast numbers of people to the intellectual and sensual pleasures of discovery, of bold new tastes and the appreciation of nuances. The respect for variety in the bottle resonates back to a respect for the land, because promoting the *terroir* of individual plots of land requires protecting the innate character of the soil and its local context. Intrigued by the variety in the bottle, wine lovers are drawn to their regions of origin. In those landscapes they find not only vineyards and tasting cellars, but also cultures and economies rejuvenated by hundreds, of competing and cooperating winegrowers.

Many other industries try to emulate the success of the wine industry. Producers of coffee, chocolate, cheese, and other agricultural goods, and some manufactured products, seek to replicate the success winegrowers have had in promoting place of origin as the fount of profitability. Although producers in many regions have made significant achievements in using this tool the underlying dynamic within wine territories is not being exploited. It is a dynamic that other alternative forms of production, such as community-supported agriculture or fair trade, should understand if they want to prevent homogenization and commodification from crushing their efforts. Within wine territories, winegrowers compete with each other, differentiating their products, while at the same time building a collective reputation and brand. Integrating what can be opposing objectives among hundreds of winegrowers is not easy. This book uses a comparison with corporate organization to reveal the dynamism and greater sophistication of territorial organization.

Territorial organization provides the economies of scale and scope needed to compete against large corporations: marketing capacities, leverage and mediation in distribution channels, R&D, political lobbying, quality and production controls, management, and most importantly the greater recognition of the territory's brand. Furthermore, the territory, its landscape, culture, and internal diversity provide an image and experience that cannot be matched by a corporation. But while winegrowers need economies of scale and scope, territorial organization cannot eclipse individual efforts at differentiation. Rather, territories must support the winegrower's differentiation efforts and help consumers sort through the variety offered by its many producers. Territories, thus, organize to achieve collective goals that recognize individual differentiation in a manner that corporations cannot. Bordeaux, Napa, and Chianti Classico, the three wine territories featured in this book, are leaders in developing these institutions.

I did not come to this research or appreciation of wine naturally. The culture I was raised in favored another beverage as an accoutrement to watching hockey games and other less important events. I must also confess to rolling my eyes when others swirled glasses. Moreover, my research as a geographer had been focused on Japan where I had focused on inter-firm relations as expressed in a number of industries. It was, however, that interest in inter-firm relations that piqued my interest when my wife's hamstring injury halted a bicycle trip and left us marooned in Bordeaux. A tour to Medoc's chateaux introduced me not only to the culture of wine, but to the remarkable and disciplined way the winegrowers of the classification and territory agreed to rank each other and their profitability. The logic of horizontal cooperation among thousands of competing chateaux was more intriguing than the vertical relations I had studied in Japan. I would learn, moreover, that the Medoc classification is but one of several mechanisms of cooperation that Bordeaux, let alone other winegrower territories, have created. One should beware of following their interest.

This book has taken 10 years of my life. The most enjoyable time in that decade was the field research in the three wine territories that make up this book and several others in Europe, South Africa, New Zealand and North America. In each of the book's three territories I conducted formal interviews with elected presidents, past presidents, directors and key administrators of winegrower associations. I interviewed representatives of several other associations and governments related to or impacting territories (e.g. chambers of commerce, organic associations, and universities in the territories and the INAO and OIV outside). The interview objectives were to determine how the association operated, and with what degree of winegrower input; what services were provided to the winegrowers; and how the various conflicts of interest in the territory are dealt with. Publications and statistics detailing policies and practices were obtained and used extensively. As I note in the text, the people who run these organizations, the winegrower directors and the professional employees are a small core of expertise at the heart of large territorial organizations. They kindly spared the time to walk me through their operations. Special thanks go to Anne Marbot and Philippe Code at the CIVB; Jean Lissague and Christophe Terrigeol at Blaye; Christophe Chateau at Côtes de Bordeaux; Hubert de Boüard and Nadine Couraud at Saint-Émilion; Dawnine Dyer, the late Tom Shelton and Linda Reiff in Napa; and Silvia Fiorentini, Giovanni Ricasoli-Firidolfi, and Emanuela Stucchi Prinetti in Chianti Classico.

In each territory I interviewed from 20 to 40 winegrowers to determine how they differentiate their products, which associations and services they used, the effectiveness of the services, and winegrowers' participation in the governance and operations of the territory. I interviewed a spectrum of winegrowers: world-renowned estates, less-renowned estates, recognized leaders in the territory, cooperatives, newcomers, ancient families, organic producers, and large volume wineries. All the winegrowers made an impression on me and on this book. It was an enlightening experience to talk to such diversity of people who share a passion and excellence for their profession, one that encompasses such a breadth of competencies. Although it's a disservice to many I interviewed, I would like to thank a number of winegrowers who represent the support I was given. They are Paolo de Marchi, Carlo Mascheroni, Sylvie Haniez, Andy Beckstoffer, Andrew Hoxsey, Greg Brown, Eric Bantegnies, Yann Boucasse, Jean-François Pommeraud, Jean-Luc Thunevin, Pierre Chatenet, and Françoise de Wilde. A special thanks is offered to the Dartier family for their kindness during our stay in Mazion and for the tractor alarm clock.

I did not limit the written research to one or a few fields of study, rather in order to construct the book's unique approach and analytical framework I sought information where I could as long as it was credible and fit or modified the logic of territorial governance. Hence, the publications, Internet sources, and other materials used are very diverse, ranging from wine guides, blogs and trade magazines to journal publications and local histories and geographies. Trade journals, local newspapers etc. were accessed daily through the use of wine industry Internet portals and newsletters. In the time that I have been studying the wine industry I have witnessed its cycles and dynamism and I believe the book reflects the territories' capacities to deal with change. That said, an academic book is only as useful as the theories it is based on. The interpretation I offer here derives from the work of four sets of scholars. Philippe Roudié and Jean-Claude Hinnewinkel provided the territory and *terroir* perspective (and the use of the CERVIN library). The territory, enterprise and regulation framework was worked out by the New Zealand-France connection of Nick Lewis, Warren Moran, Philippe Perrier-Cornet, and John Barker. The territorial differentiation thesis was constructed by Eric Giraud-Héraud, Louis-Georges Soler, Sébastien Steinmetz, Hervé Tanguy. Elinor Ostrom's analysis of the self-governance of common pool resources offered a way to bring these several ideas together.

On the ground in Bordeaux and Chianti Classico, and in cyberspace, Marie Christine and Jacques Brémont provided invaluable friendship and support when my French and Italian needed it. Leslie and Dave Proctor helped to open doors in Saint-Émilion and Blaye. A mentor, collaborator and great friend for a quarter century, Roger Hayter was called on many times to help me sort this project out. Although the project seemed to defy family tradition, my brothers and mum humoured me, even becoming (partial) converts to this culture I had joined. My greatest thanks goes to my wife and boys, who while happy participants in field excursions (a delicate balance between preferences for tasting rooms and the occasional ride on a tractor), had to live with the vicissitudes of the writing process. I promise never to put them through it again.

# List of Acronyms

| | |
|---|---|
| AO | Appellation of origin |
| AOC/AC | *Appellation d'Origine Contrôlée* |
| AVA | American Viticultural Area |
| AVF | American Vineyard Foundation |
| BAFT | Bureau of Alcohol, Tobacco, Firearms and Explosives |
| CGPA | California Grape Protection Association |
| CIVB | *Conseil Interprofessionnel du Vin de Bordeaux* |
| CNDOIGTV | *Comitato Nazionale per la Tutela e la Valorizzone delle Denominazioni di Origine e delle Indicazioni Geographica Tipiche dei Vini* |
| CPR | Common Resource Pool |
| CWA | California Wine Association |
| DO | *Denominación de origin* |
| DOC | *Denominación de origen controlada* |
| DOC | *Donominazione Di Origine Controllata* |
| DOCG | *Denominazione di Origine Controllata e Garantita* |
| ENITAB | National Agricultural Technology Engineer's College |
| FGVB | *Fédération des syndicats des grands vins de Bordeaux à Appellation Contrôlée* |
| FWC | Family Winemakers of California |
| GI | Geographical indication |
| IGT | *Indicazione Geografica Tipica* |
| INAO | *Institute National d'Appellations d'Contrôlée* |
| INRA | National Institute for Agronomic Research |
| ITV | Institute for Vine and Wine Technology |
| NSWG | Napa Sustainable Winegrowing Group |
| NVGGA | Napa Valley Grape Growers Association |
| NVV | Napa Valley Vintners (renamed from NVVA) |
| NVVA | Napa Valley Vintners Association |
| OCM | Common Organisation of the Market in Wine |
| ODG | *Organismes de défense et de gestion* |
| OIV | *Organisation Internationale de la Vigne et du Vin* |
| PDO | protected designation of origin |
| PGI | protected geographical indication |
| TRIPS | Trade-related aspects of intellectual property rights |
| TSG | traditional specialty guaranteed |
| TTB | Alcohol and Tobacco Tax and Trade Bureau |

| UGC | *Union des Grand Crus* |
| UIV | *Union Italiana Vini* |
| VQA | Vintners Quality Alliance |
| WI | Wine Institute |

# Chapter 1
# The World in a Glass

## Anomaly or Alternative?

In a relatively short length of time the corporation has become the dominant mode of productive organization. In the last 100 years especially, complex companies organizing large-scale production and managing far flung supply and distribution chains rose to dominate the global economy. These companies produce a wide assortment of goods in a range of prices. We have come to accept their organization as the norm. But are there alternatives? Could such an alternative do a better job?

The production of wine seems to be an anomaly among the world's industries because of two main reasons. First, the wine industry offers an incredible variety of products to consumers. Second, independent producers make the majority of this variety. Most independent producers are winegrowers, combining the talents of grapegrower and winemaker, they make wine from their own grapes, from their own land. The winegrowing relationship may be shared between winemakers and grapegrowers when each commits to improving the relationship between the wine and land it comes from. Evidence of this variety and its origins is on display in any neighbourhood wine shop (at least in non-oppressive countries). Hundreds, if not thousands, of varieties line the shelves, arranged according to countries, and usually by region. Except for the volume brands, the labels carry the name of the estate and winegrower that put the wine in the bottle. The wine shop represents only a selection of the vast variety of wines being bottled, but compare its variety with a shoe store or the range of any product type in a hypermarket. In either situation, a few companies with a range of brands control what's on offer. There is another less obvious reason why the industry is an anomaly. Cultivating and bottling the inherent qualities of each unique parcel of land transmits an exceptional environmental message.

The argument of this book is that this variety makes it to the shelves, not because of any inherent capacity in the winemaking process, but because of an alternative organization of production, and distribution. Although wine originates in regions, in reality, its production and distribution is often organized by territory. Territorial organization can provide economies of scale and scope that can match corporate organization. It can provide more variety to consumers, capture more value for winegrowers (aka farmers), and protect environment and social fabric. In other words, it can do a better job than corporate organization. This book provides a theoretical background for understanding territorial organization and provides three examples of how it is implemented.

In this chapter, I begin by outlining the economic foundations of these benefits, the importance of variety to consumers and to producers; and how its demand and supply are structured. An introduction to the evolution of wine variety follows. Subsequently, I compare how corporate and territorial organization provide varieties of wine. In the latter discussion, I isolate democratic self-governance of a territory as a key focus for discussion in Chapter 2. The chapter concludes with a brief rationale for choosing Bordeaux, Napa, and Chianti Classico as case studies of territorial organization.

## The Consumption and Production of Variety

According to Napa legend Warren Winarski, wine variety is composed of the contributions from the grape, the ground and the guy. The grape or blends of varieties provides the basic framework of chemical properties, growing period and conditions that provide the basis for taste characteristics. The ground consists, not only of the soil, subsoil and slope, but also the micro-climate, regional climate and biotic characteristics of the surface and subsurface ecology. The guy matches grape to ground, cares for their interaction, and the winemaking itself. The winegrower has a suite of practices and technologies capable of producing disparate outcomes from the same baskets of grapes and depending on the growing conditions of the year will blend different grapes together. Wine, thus, has the capacity to be highly variegated. The French use the term *terroir* to designate the wholistic and unique interrelationships between Winarski's three factors and use it to link vineyards, estates, and regions to the attributes of a bottle. The *terroir* definition of variety has become a universal standard for wine territories. It allows winegrowers to describe the unique features of their land, portray their own creativity, and not incidentally, justify higher prices. But can consumers absorb such variety and can it be produced and distributed efficiently by producers?

### Consumption: Variety as Equity

> ... individual variations in tastes or preferences are real and substantial in the sense that individuals consider themselves to be better off (or have higher welfare) when they have a product which exactly fits their view of the ideal design for that class of products than when they do not.[1] (Kelvin Lancaster)

Intuitively, we know that variety is a range of choices and that it is important. Among cars, houses, mobile phone providers, breakfast cereals, or tomatoes we choose a variety based on our income, the various functions it can perform, its looks, what other people think of it and that which one's spouse really likes. At its core variety arises in significant and less significant differences between products. Economics, largely, accepts the assumption that our demand is highly variegated, leaving it up to the individual to decide on the utility or value attached to characteristics attached to a products. Whether a consumer has the capacity to select an appropriate variety of product is a measure of equity within an economic system.[2] The question then is how the consumers will make their choices among products, and therefore requiring the categorization and measurement of the heterogeneity of choice. Lancaster initiated this analysis by defining product variety as a bundle of characteristics, each of which can be differentiated vertically and horizontally.[3] The composite of these characteristics also differentiates the product vertically and horizontally among its peers.

Vertical quality describes differences of degree, while horizontal describes those of kind. Bigger engines or less noise in a car designate vertical qualitative differentiation, while a range including, sedans, SUVs, coupes, or ragtops, represents horizontal differentiation. Tomatoes can generally be compared for redness, roundness, sweetness, or freshness, but plum, beefsteak, salad, or cherry tomatoes are chosen for characteristics and uses distinct from one another. Importantly, while some products such as sugar or phone-time are divisible, most products are indivisible, bought as a package of goods and services. A good is not only a composite of materials, functions, and style, but may incorporate an image, a warranty, support during use, pleasantness of sales staff and so on. Similarly services are provided using a range of physical goods that customers hope to avoid investing in and learning to use. The design of these total packages provides the potential range of different products for people to choose from.

A bottle of wine represents both vertical and horizontal differentiation. Any two red wines, or two merlots, or two same-region wines can be chemically analysed for sugar content or taste tested (organaleptic analysis) to determine which is better according to common characteristics. For a 20- to 40-year period vertical comparison of wines within a grape variety, a varietal, defined the meaning of quality in the US and created a globally powerful categorization of wines. Winemakers, critics, oenologists and the government backed this categorization because it helped to shift US viticulture away from poorer grapes to the so-called noble European varieties. Comparison of reds and whites, wines from different grape varieties, or wines from different regions is done on a horizontal standard. The real adventure begins when consumers seek to differentiate among the thousands of wines produced by individual winegrowers. Each will be differentiated according to vertical characteristics such as colour, alcohol, or fruitiness and these characteristics will be mixed in such a way to provide a unique horizontal identity in taste, bouquet and appearance. The identity is the various raspberries, mangos, truffles, violets, or *pee pee de chat* descriptors that are the staple of wine snob jokes, but also the estate where the wine came from, its history, the humour or heritage on the label, even the relationships between producer, distributor and buyer. Packaged together in a bottle distinctive to the region and with *terroir* as the glue, the composition becomes an indivisible good.

Determining if such attributes of variety are actually important to consumers is a concern for wine companies and territories alike. The value of these attributes have been confirmed by many hedonic studies that have estimated the relative contribution to wine prices of sensory evaluation (tannins, aromas, acidity, etc.), quality-price ratios (value), grape variety, vintage, region, ranking, awards, wine critic rankings, *terroir*, appellations, production factors, proprietor and estate reputation and so on.[4] The contribution of each of these attributes varies, however, with the geography and demography of the survey. Reputation, for example, seems to matter more in European countries where wines have had more time to develop renown, where US and Australian consumers depend more on critics' rankings and awards. Discrete choice analysis has also been used to determine how consumers choose among these attributes, with a recent focus on determining how appellations or region of origin compare and are influenced by brand, price, whether estate bottling, grape variety and other attributes.[5] Again, the results of these analysis show that it matters whether a population surveyed is expert or novice or whether it is in country with long wine traditions or not.[6]

If winegrowers are increasingly confident that their means of differentiation, in part or whole, satisfies some part of the market, there is less surety about the consumer's ability to sift through hundreds of appellations and thousands of labels. After all, perhaps the greatest advantage of a brand is to make choice easy. The insecurity of winegrowers is not lessened by research that shows that consumers do have difficulty.[7] Fortunately, this overload problem can be reduced if consumers are offered an ideal point or reference upon which to make their decision.[8] Moreover, offering a legible and credible range of variety increases the likelihood of a purchase,[9] even the simple categorization of varieties.[10] At any rate, companies offer a range of product variety even when markets are uncertain in order to ensure they have their bases covered.[11] Variety and pricing strategies are coupled as particular price points both present psychological barriers to spending more and induce people to make more expensive purchases. These type of practices reduce the dilemma of discerning quality differences among products for consumers and increase total demand for producers.[12] These problems are exacerbated by the fact that wine, especially *terroir* wines that change significantly by the vintage, are experience goods, whose quality can only be discerned after the bottle is opened.[13] The previous record of the territory and winegrower, their reputation therefore, become all the more important.

*Production: Variety as Efficiency*

> A car for every purse and purpose. (Alfred Sloan)

The winegrower's stimulus to produce variety is simple. His or her primary goal is to differentiate their product in the market and obtain market power. By producing their own label, sourcing grapes only from their piece of land, adding their own touch to winemaking, winegrowers can differentiate their product and add value. They may do this for love or money,[14] but in either case without differentiation, winegrowers would be price-takers. They could supplant the farmers, producing homogenous commodities such as wheat or apples, that are the standard examples used in economics textbooks. The farmer, as price taker, receives only a small proportion of the end product value, the price-taking behaviour is not imposed or expected of other companies farther down the value chain, such as logistics companies, distributors, grocery companies or marketers. Without market power the winegrower, like the farmer will be constantly exposed to downward pressure in prices paid for commodities. They would be forced to compensate by expanding production through mechanization, chemical enhancements of soil productivity, and expansion of acreage or go out of business. Collateral impacts include homogenization of crop varieties, unsustainable farming practices, elimination of family farms upstream, and crucially from the variety perspective, the loss of any distinctive qualities of the plant, land, or winegrower creativity. The damage imposed on primary producers because of their lack of market power has been examined since Kautsky's critique at the end of the 19th century.[15]

The most remarkable feature of the wine industry is this capacity for winegrowers to develop market power. They can vertically differentiate their wine from a neighbour or anyone else by ensuring the ripeness and health of their grapes, using better equipment, limiting the use of chemicals, using barrels, expending more time and care in pruning, fermenting and racking, and so on. They can horizontally differentiate in their selection of grape varieties, by matching variety to soil differences, slope and aspect, blending wines from different land parcels and vats, choosing particular equipment and processes to use, and so on. However, this capacity to differentiate rarely suffices.

Differentiation raises winegrowers from the perils of pure, commodity-based markets to the realm of monopolistic competition and the prospect of greater equity through variety.[16] Monopolistic competition, however, creates an organizational paradox.[17] Differentiation usually requires a higher than average cost, resulting in a trade-off between variety and costs. At it simplest, the organizational question is, how can equity, in terms of providing consumers with freedom of choice in their product preferences, be reconciled with efficiency of production?

Counter-intuitively, the demand for variety seems to drive us away from its efflorescence because of the constraints it places on monopolistic competition, that is, a lot of firms making a diversity of products. Even if independent firms can keep the costs of production under control, there is still the need for economies of scope in distribution, marketing, finance and management. Furthermore while large firms may adopt differentiation strategies and organizational capacities that enable them fill many segments of a product range, the provision of variety may be reversed because multi-product and product differentiated firms can use these as tools to reduce competition. Economies of scale and scope can be used to erect barriers to entry, such as product proliferation and advertising. As in the case of GM, what started out as a multi-divisional, multi-product, and myriad attribute provision of variety, not only reduced and weakened competition in the industry, but also imposed homogeneity (or only a verisimilitude variety) on its products for presumed production and distribution efficiencies. The consequences are more pronounced in agriculture-based industries, where most product categories are highly concentrated, with an oligopoly of four firms dominating and often a duopoloy or monopoly reigning at a regional level.[18] The farmer's price-taking position is reinforced as cost pressures are pushed down the supply chain.

Monopolistic competition among hundreds or thousands of producers seems as unlikely to exist in the wine industry as it in the beer or cereal industries. The economies of scale and scope available in production, distribution and R&D, not to mention barriers to entry from product proliferation and advertising strategies appear to preclude such a profusion of variety. But perhaps the assumption that it is only the prerogative of large firms derives from a lack of an alternative example. Winegrowers, using the territory as an organizational base, have solved the paradox of variety's balance between equity and efficiency. They provide an example of how a high level of monopolistic competition and variety can flourish. But in so doing, they create another paradox. This competition is achieved through collective organization.

### Two Paths from Plonk: Corporate and Territorial Organization

Wine has always had a dual identity. On the one hand it was a locally grown and consumed staple of the diet, highly heterogeneous in grape variety and blending. It provided needed calories, compared to water was relatively safe to drink, and was not entirely appreciated for its finer qualities. On the other hand, for millennia, wines grown in particular places, and with particular varieties, have been highly valued for their characteristic qualities. These particular wines were the object of long standing and long distance trade.[19] Over the last several decades the distinctions between cheap, non-descript and distinctive wines have been blurring and segmenting, driven by the shift to an urban based market and an increase in income and expectations for quality.

A new duality has evolved, based on the definition of variety, and complicated by the strategies of two different organizations of production. On the one hand is the familiar corporate form and on the other is the less recognizable territorial form. Both have the capacity to provide variety but the wines of corporations are primarily differentiated by a varietal or an allusion to a European origin (e.g. Burgundy or Chablis). The latter is a blend of undeclared grapes and called semi-generics in the US. The corporations initially awkwardly accommodated *terroir* within their brand range but now seek to capture its benefits. Territorial organization is based on *terroir*, but expressed at different spatial levels; blends from the entire territory, sub-territories within it, estates, and specific vineyards. Most importantly, the territorial organization is made up of many independent producers, and coordination of interest among them is not without its issues.

A duality between forms of production is not unique to the wine industry. Almost all industries allow for specialist firms, particularly those offering higher end products and generalist firms offering both mass-produced and specialist products.[20] Indeed the history of the wine industry in the US, Australia, and New Zealand, with a domination by a corporate oligopoly and a still proportionally minor growth in specialists suggests a similar trajectory. But bottling the expression of the land – predominately estate winegrowing – is not necessarily a traditional or niche based production. At the outset of the 20th century a small number of France's one million grapegrowers made wine and only a relative handful marketed them as a distinct product. The overwhelmingly dominant system of wine commercialization was through an interaction of independent winegrowers with merchant-blenders (Loubère 1990, 137).[21] Today, of France's 110,000 commercial grapegrowers, estate winegrowers vinify 49 percent of the total harvest, with another 45 percent of winegrowers taking their harvest to one of the 840 cooperatives, and the remaining 7 percent selling their harvest as grapes, juice or must.[22]

In the last decades of the 20th century and the first of the 21st century, estate winegrowers grew from insignificant numbers to thousands in the US, Canada, Australia, Italy, Argentina and elsewhere. With important variations, they are following the French model of not only linking vineyard with bottling value-added, but also adopting the territorial organization needed to compensate for the limitations of being a small sized purveyor of uniqueness. Producing these wines does not require prices approaching Bordeaux Grand Cru or Napa cult wines. Remarkably, consumers do not have to pay an outrageous premium for this variety because of territorial organization.

*Corporate Organization*

Corporations generate their variety in marketing departments, laboratories, and corporate boards. They base their decisions on perceptions of market demand, potential efficiencies in viticulture and vinification, and how management can accomodate the resulting range of product offerings. There is a lot that is right with this attention to the market and production efficiencies. Indeed it is the overwhelming dogma of what successful corporations should do. Wine companies following this path of differentiation are simply adopting the best practices followed in most industries.

Corporate organization brings to the wine industry the same advantages of economies of scale and scope used in most other industries. Mechanization has been brought to bear on vast acreage of vineyards. At a basic technical level, the cube-square law and the principle of multiples describe the advantages gained from producing wine in enormous vats and integrated plants. Another fixed cost advantage is the investments in R&D that transformed grapes from disparate areas and varieties into consistent, standardized and drinkable quality. The advantages of scale in production are, however, of limited use in producing a variety of *terroir* wines. The cost imperatives of blending are obliterated when a ton of grapes from a renowned vineyard costs 15–20 times that of the grapes going into an average bottle. Intensive attention is added in the vineyard and cellar. Smaller, more specialized equipment is used in viticulture and vinification. Higher quality corks, bottles, labels, and other materials are used. Each new barrel used to make

300 bottles of high-end wine will cost approximately US$1,000, thereby adding US$3 to its cost.

Still, corporate integration reduces some of these costs and enables large producers and distributors to transform fixed costs into variable costs and to transfer risk. Purchasing power provides leverage over fragmented suppliers allowing corporations to scale production up and down while forcing the costs of market and climatic fluctuations onto grapegrowers and wine producers. Integration enables greater extension into distribution and to marketing – activities where small producers are severely handicapped. At an extreme, a large firm may establish its own distribution outlets, but more typically imposes leverage over shelf space among wholesalers and retailers, and can expect more sales effort from them. Advertising outlays, while not huge in the wine industry, are significant, especially in reinforcing leverage within the marketing channels.

Coordinating activities and sending appropriate signals to consumers through price points is a key strategy. For the last decade or so price points in the wine industry have begun at basic (<US$3 per bottle), popular premium (US$3–5), premium (5–7), super-premium (7–14), and ultra-premium (14–150) and icon (>150). Brands in the first three-fifths of the range are largely devoid of geographic and even grape variety differentiation, blending together grapes brought across regional and even national borders. Differentiating characteristics accumulate, however, with basic wines identified simply by the brand and a consistent flavour and quality, while grape variety (varietal), complexity and body, origin, winemaker, and image compound and become more refined as icon status is approached. *Terroir* is introduced at the super-premium and icon levels. Integration provides synergies between the brands, where a renowned ultra-premium or icon wine or a famous winemaker build brand equity and support brands in the company's lesser price points.

The brand, and the building of brand equity, makes choice simpler for the consumer by establishing a reputation for product quality, functionality, price, and image.[23] The consumer becomes loyal to the brand because of its reliability and consequently successful brands return significant premiums. The attenuation of firm activities into brands reduces administrative costs and enables the legal defence of trademarks. There are various ways to build a brand strategy, but generally it involves a balance between an overall image (corporation, brand family, umbrella brand) with a distinct brand for each product. Most corporations develop brand portfolios that allow them to take more shelf space, offer consumers more choice, obtain administrative and marketing economies, and even incite internal competition. Firms offering such a portfolio of differentiated products reduce transaction costs in a manner attractive to wholesalers and retailers. Product proliferation is also a strategy to prevent entry into an industry.

Gallo, for example, has over 50 brands including its brandy and wine coolers. Andres, Carlo Rossi, and Livingstone are produced in large volume and low cost and without grape variety distinctions, while premiums begin to be attached to varietals such as Redrock, Redwood, and Turning Leaf. Higher prices are attached to varietals and non-descript blends that come from Californian regions or from the main wine producing countries. Examples of the latter are Red Bicyclette from France and Da Vinci from Chianti in Italy. Gallo's first foray in super and ultra-premium wines was the creation of Gallo Family wines in Sonoma, and even there a range of qualities and prices were created. The most recent shift in strategy was to purchase ultra and icon wineries in Napa such as William Hill and Louis Martini, capturing their traditions and value added. Constellation Brands covers a similar price point and geographical range with an even greater diversity of offerings. These strategies are based on extensive and sophisticated market research, perhaps guided by academic studies or employing academics, but usually deploying more resources. For example, in a study completed in 2008, Constellation surveyed 10,000 premium wine drinkers. That study found that the 23 percent of consumers who are overwhelmed by choice reduce their purchases, but also that enthusiasts (12 percent) and image seekers (20 percent) account for half of sales.[24]

Other advantages of scale and integration include depth in legal expertise and political influence. These are not trivial in an industry subject to a plethora of health, environmental, labour and other regulatory issues, where trademark protection is critical and that has critical interests to protect at several levels of government. The importance of administration and marketing is reflected by its doubling of the costs of a bottle, and the potential synergies of integration reflected in the fact that mass production bottle costs are dramatically lower. But still wholesaling and retailing double the cost wine to the consumer, and lay open a further potential for integration, the benefits of which are not captured just by large wine companies.

Large supermarkets in Europe and the US are increasingly deciding what variety wine consumers are offered. Supermarkets, already oligopolistic for retail sales of groceries in developed country markets,[25] are taking over wine sales. Their influence is strongest in Northern European countries (e.g. UK 50 percent, Netherlands 62 percent)[26] that have little domestic production and are important markets. Retail integration is also important in North American food services because of the dominant role of chain restaurants and their preference for large volume brands.[27] These national and regional oligopolies, in turn are monopsonies and can impose their terms on the supply chain. They have reduced supplier numbers, while demanding more in terms of services and a greater range of products. This interplay is animated by a struggle over margins, while it selects and strengthens some suppliers for their ability to invest in innovation, distribution, and marketing. The supermarkets are also enticed to move into the supply chain, cutting out the middlemen and developing their own private brands. In the UK, supermarket private brands account for 50 percent of their sales.[28] The benefits of integration, particularly those related to marketing and the capture of value-added downstream in the value chain, thus are resulting in a dual oligopolization of the global wine industry. One stop shopping at Tesco, Carrefour or Trader Joes is being matched with multinational producer-distributor firms increasing scale and scope through acquisitions to exploit the globe's vineyards and wine markets.[29]

Using strategies that belie characterizations of New/Old world rivalry, European, North American and Australian multinationals are acquiring brands across the globe for their product lines to meet the needs of the giant distributors.

Perhaps the most taken-for-granted but critical aspect of integration is that it occurs under a relatively simple form of hierarchic authoritarian control. While the market is used to minimize the costs of downstream commodity inputs, ownership of critical upstream production, distribution, the brand, of the organization allows rapid and effective coordination of its critical value added components. The differentiation strategy, its impact up and down the value chain can and is changed by fiat on a frequent basis. Authority over differentiation is crucial, because ultimately it is differentiation that adds value and legitimizes capturing the profits from value added.

The differentiation strategy used by the dual-oligopoly is not, however, completely of their own making. The wine producers and distributors are capturing a source of differentiation generated by the *terroir* and estate winegrowing traditions. They created the basis for distinguishing each wine by its origin, creativity of winegrowing, and selectivity of grape varieties. As growth in mass consumption wines declined, and as growth in distinctive better quality wine grew, large firms shifted into the production and selling of *terroir* differentiated wines. For example, by 2000 in the US, the 20 percent of wine sold at $7 or more per bottle accounted for 43 percent of the value of wine, having tripled from 7 to 20 percent in volume between 1980 and 2000. Some such as Gallo made the shift by developing their own *terroir* based wines, others bought out established estates. Constellation bought dozens of estates and smaller brands and has even gone so far as to divest itself of its 'jug brands' to concentrate on anything that can carry the cache of a bottle. Yet this is not the dominant means of organizing the production of variety.

*Territorial Organization*

Since ancient times particular places have produced distinct wines and generated the complementarities that propel trade.[30] The relatively cool Nile delta in Egypt, Chios, Lesbos and Thasos in Greece, Falernum and several other "Grand Crus" of Rome are examples of these unique places.[31] Commerce of course had to be organized – the transportation to markets, distribution and sales, and the production within the wine region. The organization of wine producers into a group and territory interested in preserving and developing the reputation of the region's wine is thus part of that long history. Saint-Émilion is perhaps the best-recorded example of that experience. Yet, only in the last 200 years have territories evolved to represent not only a distinctive wine, but a collective of many winegrowers each making distinctive wines. And only during the last half-century territories have developed as sophisticated platforms for the generation and organization of variety.

The modern wine territory offers both a territorial reputation to its producers and enables them to sell a distinct product. Thus while a wine territory usually produces a blend or generic wine drawn from producers selling all or part of their grapes, must or wine to wineries, the territory must incorporate a substantial proportion of estate wineries or other means of designating *terroir*. Each of these estates or vineyards must sell some proportion of their wine or grapes under their own brand. Predominately, wine territories are distinguished by estates, but more essentially, the capacity to allow *terroir* and winegrower creativity to be expressed in diversity.

A territory's disposition to variety inhibits the large economies of scale that are the pillars of corporate integration. At the production level, each winegrower chooses their own quality of corks, their own labels, or a particular toasting of their barrels. They need their own equipment because their neighbours' equipment will be put to the same use at about the same time. Nor are economies of scale easy to obtain in distribution and marketing. The independence of each winegrower, the marketing of their own brand, inhibits collaboration. Yet, it is where scale economies irregularly blend into scope economies that winegrowers have creatively integrated their territories creating organization that in scope, inclusivity and flexibility can be more sophisticated than the modern corporation. For the corporation, the brand or brand family provides the consumer with a signal of a product's quality and image. The success of a territory depends on establishing a reputation that can assist the brands of its winegrowers.

The importance of collective reputation has been confirmed in several empirical studies linking prices to the territorial reputation[32] and to consumers placing greater confidence and paying higher prices for territories that have proven their qualities over the long term rather than in a short-term performance.[33] Brand and place of origin concepts are being integrated to provide a nuanced signal.[34] The primary goal is to create a system to overcome variety overload and the difficulties of ascertaining quality within the territory's diverse wines. Consumers search for reputation as defined by a territory's unique set of characteristics (e.g. grape varieties, physical attributes, viticultural and winemaking practices), to avoid a costly search to assess many individual wines. For example, many wine territories, through evolution or design, focus on one or a few types of wine, not only because of the complementarities between grapes, ground and traditions, but to send a coherent signal to consumers. The same logic holds for the reputations for winegrowers within a territory. Territorial reputation becomes a trade-off between providing a coherent signal to consumers about the territory's qualities and allowing for a differentiation of offerings from independent winegrowers.[35] Without this layering of reputations, information asymmetries between consumer and producer spawn a destructive spiral where consumers can't tell good from bad and refuse to pay higher prices for better wine.[36] Quality producers lose the incentive to invest and consumers lose a range of products. Within the territorial reputation, winegrowers are still able to differentiate their wine and respond to consumers' demand for variety. Many territories devise classifications, appellations and other means to assist choice.

National and international trademark laws recognize the value of using territory as a brand, as deserving of the monopoly power and legal protection provided to corporate brands. These trademarks are known as appellation of origin or geographical indicators.[37] In a world that views the corporation as the convention, the protections given to territorial rights are more ambiguous and receive less respect than corporate trademarks,[38] but there is a more significant weakness. Corporate trademarks are owned privately, are transferable, and can be authoritatively protected and supported. On the other hand, the appellation is a non-transferable public good, shared by producers in a region and requiring national regulations for creation and existence. The legal framework may also provide producers the power to put controls on who can use the appellation and how they can use it. Appellation laws do not, however, recognize the contributions of independent producers to the territorial reputation or provide for proportional compensation. The inability to transfer the regional trademark is partially compensated for by the transferability of land that has had its value increase by territorial reputation, but the winegrower's incentives to develop their own brand remain large. The interrelations between territory and brand can result in tensions, particularly when one estate believes it has built the reputation that others in the territory are using.

The need to accommodate independent and competing winegrowers within the same territorial organization is the pre-eminent difference between it and the corporate organization. As a consequence the two differ in their governance and coordination requirements.

The territory is not governed through the singular authority afforded by sole ownership, partnership, or management for stockholders. Wine territories are governed collectively, typically democracies of property owners who vote directly on significant issues and who elect an executive to guide and administer the territorial organization. These organizations are akin to business or trade associations, but are tied to and influenced by their locality.[39] The organization is part of the actual territory it inhabits, and it exerts an economic and cultural force to generate the identity of the territory,[40] which to a great extent is shared by other inhabitants, businesses and government. Although territories are the creation of winegrowers, it is their national governments that provide them with legal designations and the right to a trademark, and as in France and Italy, the legal mandate to self-governance by members.

Even with government support, achieving some measure of agreement among hundreds, and in Bordeaux's case, thousands of owners of greatly differing capacities makes management of a territory inherently challenging. Two concerns are predominant. First, the actual physical extent and the membership of a territory must be delimited and maintained. This is the demarcation of their trademark monopoly and who gets to use it. Second, winegrowers must ensure that all individuals support the territorial reputation; by maintaining wine quality, not engaging in overproduction or collective investments. For both these issues, rules have to be created and enforced, supportive mechanisms and administrative organs established, otherwise the reputation of the territory will be open to free riding and degradation.

The coordination demands on the territorial organization are more complex than its corporate counterpart, and require constant referral to the parameters of governance.

Quality control is the clearest example of these, revealing both the collective need and its contradictions. Whenever a consumer drinks a glass of wine, the territory's reputation is at stake. How can the territory ensure that the quality of wine will preserve its reputation while allowing the wine to be different? A corporation can impose strict standards on its brands – make them the same vat after vat, year after year. Such standardization is anathema to the high proportion of many winegrowers who have entered the industry because of their love of winemaking[41] and it is also difficult to impose over wide disparities in investment, *terroir*, and winemaking capacities.[42] The importance of quality control is underscored by the long process of building reputation and the speed with which it can be destroyed.[43]

The difficulty of QC is underscored by the need for effective standards, regulations, monitoring and enforcement to be imposed on independent principles to prevent free riding.[44] In Europe, the massive shift of commodity wine producers into *terroir*-based territories required the imposition of extensive and intensive quality control, with widely varying results. Control on yields, vineyard and winery practices are some of the methods used.[45] Imposing quality control has less traction in the US or many New World territories, and not simply as an adverse reaction. Generally, these territories are smaller in winegrower numbers, and don't have the power to impose QC. Moreover, they can often rely on well-heeled entrants to push quality standards higher. Closely related to quality control, all the territories undertake significant collective R&D to understand plant diseases, develop new vinification methods, and other economies. Often this research is dependent on and coordinated with regional or national universities and other institutions. The same institutions and a diversity of others are necessary to diffuse the knowledge through training programs, seminars, publications and so on. This research and training must, however, be aligned with the needs of the territory and of the differences among the winegrowers and their properties, and somehow it must be paid for. Once again independent firms have to be asked to contribute to a collective cause.

*Terroir*-based winegrowers require coordination over distribution and marketing channels. Whereas corporations have the benefits of scale that allow them to control the supply chain from vineyard to market, small producers lack internal integration and the power to influence their channels. They have to look for other means. The most obvious of these is direct, and many winegrowers, family, occasionally employees, spend weeks on the road paying attention to old customers and hoping for word-of-mouth effects. Yet, direct sales rarely suffice and winegrowers must seek other channels. Usually small producers need a large geographic market and many distributors to ensure that their wine will find customers that appreciate their type of variety. They work with distributors of various capabilities and sizes, but within these relationships there are many difficulties – exclusive agents may be too complacent, while supermarket chains drive down prices, and government regulations preclude sales and drive transaction costs up.

Variety lies at the heart of the tensions within the downstream value chain. The proliferation of estates and their brands decreases marketing incentives for merchants who have neither capacity nor incentive to represent dozens of chateaux adequately.[46] Rather they are likely to free ride (rely) on reputation of chateau and appellation. Furthermore, as in any buy-sell relationship, substantial friction arises among the various parties that can make up the value chain. In the wine industry grapegrowers can sell their grapes or must to wineries that don't have their own grape supply, to estates with grape supply, to cooperatives, on the spot market, or consolidators, while winegrowers can sell their wine to merchant-blenders (*négociant*), wholesalers, merchants, directly to supermarkets, restaurants, or specialist retail. Within such arrangements, the parties can be competitors or collaborators, or both. There is always potential for opportunism and strategizing that can lead to under-investments in viticulture, viniculture, and marketing, especially when the conditions of agriculture are considered.[47]

France and a few other countries have established institutions composed of stakeholders in the distribution chain that try to ameliorate its tensions. They invented several mediation mechanisms that, for example, provide price and market information for both sides or stabilize contractual arrangements. The US has produced similar mechanisms through negotiations among institutions and stakeholders. A further impact resulting from the limitations to distributor incentives and representation is to make winegrowers and territories more responsible for their reputations. An awareness of distribution constraints stimulates collective marketing. Territories promote themselves and their winegrowers at supermarkets, trade fairs, wine competitions and festivals, and other events. But the turn from dependence on marketing channels to collective marketing is not without complications. It is difficult for a winegrower to connect the impacts and costs associated with collective marketing to individual returns, particularly when producers are working hard to establish their own brands.[48] A dramatic mediation of the winegrower-buyer-collective relationship is the futures market, whereby winegrowers have reversed the speculation that normally takes place at the end of the value chain and capture value-added for themselves.

The greatest tool at the small producers' disposal and where their collective efforts prove more than a match for a corporation, is the territory itself. It draws people in, generating immediate sales and long-term customers. The creation of such a territory is a both an individual and collective act, the creation of positive externalities combined with collective investments in symbols and amenities.[49] Again tensions are likely to rise in regard to who makes the investments and who is along for a free ride. If these tensions can be overcome, however, the territory can be a highly diverse, authentic and quirk filled attraction, one very difficult for a corporation to produce. Napa is California's second most popular tourist destination, but it has too many oddballs and iconoclasts to be a Disneyland.

Territorial organization has other advantages missing in the corporation. Estate winegrowers are driven to excel and bring a passion and intensity that cannot be replicated in a company.[50] Instead of brand managers trying to determine what should go into each price point, how to prevent cannibalization or how to defend against other firm entry, territorial reputation actually fosters open competition among winegrowers. If they fail, they won't take the whole territory with them and it won't be a brand manager's rationalization strategy that terminates them.

Territorial organization performs most of the integrating and coordinating activities that a large company. It does so despite a lack of authoritative coordination and the independence of competing winegrowers that make territorial organization difficult to initiate and improve in a very competitive global market. The power of territorial organization is attested to by the fact that in many cases corporations have had to integrate their activities with those of the territories. To capture the premiums that consumers are willing to pay for variety, corporations have adopted not only the *terroir* strategies of the estates, but have integrated into territories and many proactively support them.

### Contributions and Structure of the Book

Excellent wine comes from territories. Of the few hundred wines that earn very high prices, only a couple of wines are blended from different areas. Of the few thousand wines that gain the attention of wine writers, omitting those that shill for supermarkets, the vast majority come from independent winegrowers in territories. The appreciation of wine has been at the vanguard of a movement rediscovering and creating variety latent in agricultural products and many other goods and services. Moreover, wine appreciation has shown how consumption, production, and sustainable practices can coalesce in places of production, and can project their influence into urban markets. But territories are much more than regions or origins. They are politically delimited areas, self-governed by independent winegrowers, for their collective and individual enterprise. The capacity to produce excellent wine springs from a dynamic of competition among them, cooperation on collective goals and for the imposition of controls to build and maintain the territory's reputation.[51] The objective of this book is to explore the many configurations and ramifications that result from this dynamic, but which are not sufficiently explained in existing publications.

Independent winegrowers are celebrated in a great number of popular books and academic studies. Indeed the threat to winegrower expression and diversity has been a cause célèbre of several books and films.[52] I have availed myself of their expertise extensively and hope that the analysis offered here will be useful in return. The primary audience, however, are those concerned with the theory and practice of regional and territorial organization, or simply organization. In particular the research contributes to the critique of globalization of agriculture and the exploration of alternative organizations.[53] This research has investigated the weakening terms of trade imposed by commodity chains and the consequences for cultures and ecologies. The alternative organizations include familiar local forms such as farm shops, roadside stalls, 'pick-your-own' programmes, farmers markets, and local shops.[54] This book addresses the more complex alternatives such as quality labels, fair trade, geographical indications (GIs) and particularly appellations of origin (AOs).

France's *appellation d'origine controlée* system generated protected denomination of origin and other collective quality marks in dozens of countries. Subsequently the EU has developed designation of origin (PDO), protected geographical indications (PGI), and traditional specialty guaranteed (TSG) systems and provide institutional and monetary resources to their adoption. These ideas have been supported by international trade regulations. Behind these developments is the idea that wine appellation of origin is the exemplar of how to add value to commoditized farm products. The beneficial recognition of this collective trademark, however, does not incorporate the key element generating an appellation's value added – the collective-independent entrepreneur dynamic of territorial organization. A full explanation of this dynamic and its institutional mechanisms is also missing from academic analysis of appellations and their territories. [55]

Similarly, the appellation and the broader alternative food system literature use quality as the foundation of value added but without a substantive definition and without incorporating or pursuing the concepts of horizontal and vertical differentiation. Quality, as conceptualized in the literature, is vertical differentiation, and like a varietal wine, remains open to commodification and appropriation by corporate organization. The exemplar of this commodification process is organic production.[56] Where once small independent farmers engaged in mixed farming were able to use "organic" as a source of differentiation, large corporations have expropriated the term and apply the same label to monoculture crops. Fair trade products face a similar competitive fate, as exemplified by the purchase of Black's Chocolate by Cadbury's, and other chocolate and coffee independents swallowed by large companies.

An example that indicates how a territory is a collaboration among independent and competing producers, and not simply a cooperation on a vision of territory, is that in any prestigious wine territory outstanding estates are extremely important to the territorial reputation but the territory is larger than any estate. Outstanding estates in Bordeaux are known as the "locomotives." Their formal designation as "grand cru" singles them out as the top of a very big mountain. Competing winegrowers, in spite of the prestige and value disparities created, support the system because the locomotives pull the territory and collective with them.

In the next chapter I explain the complexities of territorial formation and collective governance among competing estates. This is crucial not only for understanding internal dynamics, but also how winegrowers and their territory influence the other governance structures that shape the value chain and markets. A focus of the book is how self-governance of the territory is created and maintained and how it is used to support winegrowers' control value through the distribution chain. The next chapters discuss the organizations in three different territories to demonstrate the similarities and differences of their organizations.

Why Bordeaux, Napa and Chianti Classico? Their reputation precedes them, but these regions were built with great effort, and sophisticated organization goes along way to explaining their consistent success. Bordeaux is generator or propagator of many innovations taken for granted in the wine industry: the chateau (estate), appellations, *terroir*, estate bottling, differentiated futures markets. It's also undergone a massive transformation from bulk to estate wine sales. It has a multi-layered organization that must deal with the differentiation of over 5,000 chateaux among a total of 10,000. I spend one chapter discussing the overall organization of Bordeaux, and in order to focus on territorial dynamics and their differences, one chapter discussing Saint-Émilion and Blaye.

Napa spearheads the diffusion of chateaux and *terroir* to the New World, innovating to do so. Lacking many of the systemic attributes that support territories in Europe, it has to forge its own way within a culture and legal environment biased toward corporate organization. Napa has also developed its own forms of variety and institutional mechanisms. Chianti Classico represents the diffusion of estates, but with a twist. To establish identity within the internationalization of wine styles it is re-establishing territorial unity and control. The three regions demonstrate what type of markets, institutions and economies of scale can be constructed to build regional organization and how it is shaped by the tensions between collectivity and differentiation.

Despite their different backgrounds, and social and legal contexts, Bordeaux, Napa and Chianti Classico face similar challenges to territorial organization. As a basis they have had to define the boundaries of their territory and create policies and regulations that improve the wine and promote the territory, and these have to be acceptable to the winegrowers. The territories have to fight to establish and defend the self-governance that enables them to run their territories efficiently. They have to establish regulations for trademark protection or conditions for the use of their appellation, modify regulations and to legally defend against firms that transgress those regulations. Internally, they have to raise the quality of the wine and create a reputation that can compete on the global market. Winegrowers and grapegrowers, once content to sell an undifferentiated bulk product have not only lifted their own game but have collaborated to ensure that their neighbours are able to, or must do as well. A foundation for variety in all territories is the capacity to not only produce a diversity of wines but to be able to get them to market. Each territory has come up with its own solutions for converting the supply chain of large corporations into distribution chains for independent producers. But even in so doing the potential for the brand to eclipse *terroir* remains, driven by the temptation of growth by producers in the territory and by external firms that want to capitalize on estate and territory reputations. These and other issues confront the producers and their territories as they attempt to convey the unique characteristics of their wines, from the vineyard to the poured glass.

# Chapter 2
# Territorial Governance

The distinctive feature of territorial governance in the wine industry is that it is created and sustained by winegrowers acting individually and collectively. Winegrowers establish self-governing institutions that define the territory and its image. They impose and enforce rules, often stipulating quality controls and wine styles. Winegrower associations are the core territorial organization and have the dual purpose of improving the competitiveness of the territory as a whole and of independent winegrowers. They have to find means to make those purposes synergistic or at least compatible.

In comparison a corporation imposes management over its various divisions, brands, and employees through the centralized and legal power provided by ownership. There may be principle–agent problems between owners and managers, and the troops have to regularly be rallied with the latest version of reengineering, six sigma, or in the words of Reinhard Bendix some means to "enlist the cooperation of the many with the few."[1] Generally employees have to do what they're told. Territories, however, cannot impose management on their constituents. Yet, they face similar conditions in the global market place, demand for reputation, quality, and distribution. If they want to meet these needs similar economies of scale and scope and integration are necessary.

Winegrowers are the key territorial actors, but other local and regional stakeholders and organizations may be enlisted to define and support the territory. Contrarily, territory builders may face opposition to their vision and practices. In addition to the local construction of self-governance, winegrower associations are set within a framework of regulatory and value chain governance that shapes their business fortunes and which they need to influence.[2] Government regulation is the obvious external governance. It authorizes direct powers of self-governance and control of reputation to the associations. Government also produces a myriad of direct and direct rules, taxes, subsidies and other complexities for the winegrowers and their associations to come to terms with. Less obviously, but of no less importance, the value chain is governed by formal and informal controls over transactions: direct negotiations among firms, mediation and mechanisms within and among industry associations, and also by government arbitration and regulation. Territories need to participate in these discussions to improve their terms of trade.

The self-governance of the territory, regulatory and value chain governance are intimately related. The most notable cause of integration is the appellation. To be granted an AOC, AVA, DOC, or VQA is the reason for collective action as these designations attract merchants and consumers, and bring higher prices to the territory's wines. There are of course all sorts of ways territorial governance is interrelated with regulatory and value chain governance as a territory tries to make these realms work for its winegrowers. In this chapter I will make this interaction legible by first looking at the mechanisms of self-governance of the territory and then investigate its relationship with the other two realms. In each case there are several layers to look at, with contrasting and complementary institutions, and with contextual differences. The overarching goal is to make the external governance systems work for the territory's winegrowers, this complexity not only challenges the winegrower associations, but also our understanding of organization.

## From Water to Wine: A Theory of Differentiated Common Pool Resources

Who will guard the exclusive use of the appellation, respect for using local grapes, quality of the grapes, better than the territory's winegrowers? By extension, who will foster the marketing, R&D, or the public relations necessary to the territories reputation better than the territory's winegrowers? The central issue for a territory is the management of its collectively created resource. The territory is a vehicle designed to overcome the signalling problem, how that signal is shaped needs constant attention and development. It also needs monitoring and enforcement to avoid free-riding and opportunism arising internally or attempts from outside to use or abuse its reputation.

### Rules for Collective and Individual Reputations

Territories are usually defined in political terms, areas with some measure of self-governance over a range of internal resources and activities. Winegrowing territories focus on the particular activity of making wine and thus, while often conforming to political boundaries, wine territories may be cut out from political or economic regions. The winegrowing territory is pulled together by independent and formally competing businesses for a collective purpose. To understand how territories can be made to work, we need to understand how a single economic focus can be made the basis of a territory and collective governance by its businesses. The most advanced theory available for understanding how independent and competing businesses construct cooperation within a territory is common pool resource theory (CPR). It has been used in dozens of contexts to explain how people dependent on and competing for freely accessed water, forests, grazing land, fisheries, and other resources can construct rules that can stop destruction of the commons. They establish a governance that is not of the corporate or government type. The theory can be adopted for its insights on territorial self-governance by adapting it to the needs of differentiating winegrowers.

Winegrowers and other types of CPR users, such as fishermen or water users, are drawn together as an interdependent group by similar forces. Each group is formed through mutual dependence on a particular place. For winegrowers that may be a valley, a stretch of mountain slope, a river basin, a coastline or some other demarcated area. However, that area is only demarcated as a territory when winegrowers find a means to exclude other people from its use and create rules to make sure members do not degrade the resource and each winegrower's ability to use it.[3] To achieve these ends winegrowers must construct and enforce rules that stop people from abusing territorial reputation with an efficiency relative to the capacities of the winegrowers. To preserve territorial reputation each winegrower must maintain at least a minimal vertical quality, resist the temptation to overproduce, or to engage in any other damaging activity. There is a need to build the reputation by investing in common projects and promotions and in individual activities that produce positive externalities. To achieve these ends, self-governance must produce rules to grant rights and impose responsibilities. Without these rules, the resource – the reputation of the territory – is open to free riding and degradation.

How then do competitors, supplemented by government or other facilitators, create the norms, rules, and organizations that overcome these dilemmas and maintain a commons? How can they do so without burdening winegrowers with further costs? The relevant issues are discussed in institutional economics, industrial organization, regional development and economic geography: free riding, information asymmetries, monitoring and enforcement of rules, repeated interactions, credible commitments, trust formation, and the conventions that impart both social capital and lock-in. They must, however, be contextualized by territory.[4] A territory often creates and recreates itself by the majority of winegrowers and their offspring living in the area for a long time. They depend on their estate and its relation to the territory and do not heavily discount the future. Groups that make up territories tend to be small and homogenous and their numbers stable.

Such conditions do not preclude dynamism in the territory, but participants are able to develop norms of reciprocity and trust such that initial social capital can evolve coherently with further change. For such common values and trust to develop depends on accurate, low cost information about benefits and costs of managing the territory's reputation and of accurate monitoring and sanctioning of its rules. Investment initiatives or rule-making by the association must not impose high transaction or deprivation costs. The management of the association must also be politically legitimate. Rule-making has to avoid the extremes of unanimity or control by a bare majority, and there has to be a reasonably common understanding of potential benefits and risks associated with maintaining status quo versus change of rules.[5]

Winegrowers must make their rules according to their needs, local conditions, and financial capacities. They can impose them, monitor and enforce the rules themselves, or they can enlist state support for this governance. They create an alternative organization capable of integrating the activities of their constituents and fulfil the CPR need for integration without the imposition of concentrated private or public control.[6] To make CPR applicable to wine regions, however, it must take into consideration not only that the territory is the collective action of independent winegrowers, but also that they are producing differentiated products, rather than one uniform commodity.

With a collective reputation winegrowers are not forced to standardize their product through corporate or cooperative organizations. Such a strategy could overcome the signalling problem, allow for economies of scale in production and distribution, control volumes, and provide unified market power. It would, however, deprive winegrowers of individual market power, incentive to produce variety – and for most, their reason to be in the business.[7] This differentiation, therefore, distinguishes the collective action of estate winegrowers with that of cooperatives, which developed by blending the grapes or wine into a single brand (although due to the influence of estates, coops now strive to provide differentiation incentives and outlets).

The winegrowers in a territory differentiate among themselves by use of different grapes or proportions in their blends, by the mix of equipment and practices, in their choice of barrels, by the use of different parcels of land in different vintages, in their marketing strategies, who they collaborate with, their architecture, in their choice of dog, and most importantly through *terroir*. Respect for *terroir* is fundamental to self-governance because without it territory and estate differentiation are lost. *Terroir* epitomizes the complexity of governance because, even in the renowned territories discussed in this book, winegrowers were not born with this perspective. *Terroir* is learned through cultural immersion, conscious practice, and the support of winegrower organizations. Although the concept remains contentious, today the great majority of estate winegrowers, accept its validity and utilize its credibility for differentiation. Indeed, many present-day proselytizers dismissed *terroir* in their recent past.

Winegrowers are competing with their neighbours, with counterparts in other territories and with multi-national companies. They need their territory to set up mechanisms that recognize and support their efforts at differentiation – classification, awards, promotions, and competitions. Indeed, some sort of competition among winegrowers is necessary to propel innovation and advance in the territory. Without this competiton, there would be a demand side cost as well. In Lancastrian terms, conformity would increase, reducing variety and welfare for the consumer.

*Rivalry or Non-rivalrous?*

Territorial self-governance, because it fosters an amalgam of shared traditions and trust among neighbours and transaction partners, of interactions with formal institutions and ecological conditions, complements other approaches used in the analysis of wine appellations and territories. These include the conventions approach,[8] historical analysis of wine territory development,[9] actor-network theory,[10] and investigations of the creation of symbolic landscapes.[11] These approaches help to reveal the conditions for governance, how it can also take place unsaid and undirected, based on a history of knowledge and practices – particular ways of training vines, how negotiations take place, or even the consensus view of the status of areas or estates.

The value of these shared local practices is such that history has become a foundation for recognizing appellations not only in Europe but also in the US and other new world producers. Without such governance, the territory's differentiation is diminished. The collective creation of these localized traditions and imagery[12] is essential to the territorial project and requires that their negotiation be considered in any explanation of self-governance. Differentiated CPR analysis adds to these approaches by explicating the tensions within a territory and revealing the mechanisms used to overcome them. Similarly, this adaptation is akin to the public goods described in club theory,[13] however the non-rivalrous membership of the club does not invoke the same needs for coordination required in a winegrowing territory,[14] and this differentiated CPR approach explains how coordination can be achieved.

Self-governance is implicit in many investigations of wine territories, describing the need for negotiations in the creation of conventions, patrimonialization, and imagery, particularly in relation to the creation of AOCs. Several studies implicitly recognize the need for self-governance when describing collective activities such as quality control, R&D, marketing[15] or how when larger and smaller producers conflict.[16] The benefits of self-governance are underscored by the territories described in this book when contrasted against the lack of similar capacity for promoting knowledge diffusion and enhancing territorial competitiveness in other territories.[17] How and with what mechanisms collective actions are carried out, however, needs to be clarified, in order to understand their internal acceptance and efficiency and to understand their external effectiveness. Furthermore as territory, external regulation, and the distribution chain governance are interdependent, the basis of internal governance becomes all the more important. Animating the entire dynamic of course is the fact that the self-governance is not simply for the common goals but must facilitate differentiation by each winegrower.

## Territorial Governance

The integration and recognition of territories occurs through formal organizations, and the culture they instil within the area. Winegrower associations are the most important institutions in wine territories as they have the impetus to create a reputation and to put a boundary around it. The structure of winegrower organizations varies with the history of the area, and the relations among grapegrowers, wineries, cooperatives, merchants and estates. Estate bottlers who grow and vinify their grapes in situ have the deepest commitment to territorial association, but grapegrowers, bulk wine producers, and wineries that buy their grapes also realize the benefits of territorializing their sales and purchases.

*Winegrower Associations*

The French *syndicats*, particularly those of the more renowned appellations, were the original model of an estate-based winegrower association. The chateaux are the core of territory, providing its best-known brands and providing leadership in quality expectations.[18] The associations are generally inclusive, enjoying membership from bulk producers and cooperatives. In comparison, Napa has separate organizations for grapegrowers and winemakers, reflecting a historical, and occasionally fractious, division of labour, but the recognition of territorial reputation and the growth of estates has drawn the two closer together. Napa's experience provides a reverse guide to other New World territories insofar as most new territories are more recently developing out of a grapegrower–winery divide and many are foregoing separate entities to form inclusive winegrower associations.

A territory may have a few levels of winegrower associations, the power at each depending on unique histories and the regulatory structures. Chianti Classico is formally part of the Chianti region, but eschews association with those they believe arrogated the Chianti name. On the other hand, the Napa Valley Vintner Association initiated laws to protect the primacy of the core appellation as sub-appellations and their associations developed. Although the Bordeaux name is pre-eminent among 34 federated territories and sub-federations, winegrowers work closest with their specific associations (Saint-Émilion, for example). They do so partially because French laws give the local territory the most powerful mandate.

The primary self-governance characteristic is that winegrower associations are voluntary and democratic institutions with a high degree of direct participation and representation. Typically a president and executive are elected to deal with day-to-day issues and formalities; committees set policy; and all members vote on important issues. The democratic nature of the association, especially in relation to territorial inclusiveness, varies. As in Napa, some associations only represent their respective side of the supply chain. Similarly, some territories give an entire cooperative only one voting membership, while in others every winegrower gets a voice. Larger operations and greater status can also create power differences in some associations, formally and informally. A territory does not have to be represented by one voice, and depending on the legal situation, some territories are represented by fractious or competing organizations. Nor do all winegrowers in a territory have to be members, but getting the great majority of winegrowers to join the association is critical to legitimacy, power, and efficiency.

The policies put into practice by the winegrower associations can be divided into those backed by legal authority and those that are voluntary. All territories are a mix of the two but European (and their South American reflections) receive greater legal mandate to control all those using the appellation. American and Australian territories depend more on voluntary acceptance of collective action. In the former, the winegrower association is a precondition for establishing appellation and must be maintained to govern the appellation. A sustained self-governing institution is therefore required and provides the territory with a legal basis of self-governance. In the latter, the initial parties that establish an appellation can disband after its establishment. There is no requirement that appellation status lead to territorial governance. However, in older territories such as Napa, it was the pre-existing winegrower association that sought the appellation as a a legal recognition of their existing reputation. In newer territories, winegrowers organizing to establish the appellation usually become a self-sustaining association.

*Winegrower Association Activities*

Boundary setting is the initial and most crucial action by the winegrowers, and usually is a messy task. Historically the creation of appellations has been highly contested, not only in defining who will be included and excluded, but also by parties who may not want to see an appellation created. The consequences of exclusion from an appellation can be large for a winegrower who can no longer rely on its reputation to sell its wine. The creation of too many appellations or too large of appellations may be resisted for decreasing the significance of the system. External parties may not want to see the higher quality grapes they have used for their brands escalate in price or be removed from availability because of the appellation's impact. The formal initiation of all three appellations described in this book, despite their long histories, occurred only after significant battles and the generation of good and bad will.

After creation, the characteristics of an appellation's wine are controlled in all national appellation systems. European winegrowers may impose many obligations, not only on the members of the association, but also anybody who uses the appellation (e.g. yield limits, grape varieties, specific viticulture and winemaking techniques). Respect for those obligations requires monitoring and enforcement mechanisms to be set up and paid for by the users of the appellation. In the US controls are limited to the essential need to fill the bottle with wine made from a certain percentage of grapes from the area (85 percent generally, but 100 percent for estate wineries). A great contrast is often made between the American and European styles of appellation because of the stronger controls imposed on the latter.[19] This perceived difference is, however, based more on ideology, invidious distinction, and marketing. In all jurisdictions the differences between formally "controlled" appellations and the less controlled version are not as important as the added stimulus that appellation recognition gives to territorial association.

The controls in French or Italian appellations are only a part of association activities. Marketing, tourism, training, R&D, environmental and cultural issues are of increasing territorial focus. Furthermore, whether the controls are antiquated or not is primarily a question of innovativeness and efficiency. All of these roles, and the requisite associationalism, are driven by the need to differentiate the territory, and the realization by individual winegrowers of the linkage to their own success at differentiation. Abstract the controls, and American appellations and winegrower associations are very similar. Indeed, Napa's evolution toward a territorial identity based on cabernet sauvignon, a strong tradition of formal and informal technical exchange and recurrent recourse to the county and state governments to get authority to impose common conditions on all winegrowers, testifies to the impetus created by territorial differentiation and appellation.

The trick in any territory is not to let formal or informal controls stultify differentiation among the winegrowers. For example, controls for appellation *typicity* that penalize outstanding and innovative producers while rewarding the complacent have long been a bugbear in France. However, the extent of the problem depended on how the controls were implemented in each territory. Still, the process is being reformed. The real basis for use or lack of controls is historical conditions. The French and Italians imposed controls because they were converting pre-existing poor quality vines and peasant *vignerons* to AOC expectations. In the US most appellations spring from Greenfield sites and an influx of talent and money.

In light of this common territorial incentive, the characterization of Europeans as locked into antiquated regulations and unable to enjoy the freedom and benefits of the Americans or Australians, should therefore be viewed sceptically. Australian brand producers, blending across state lines, employ this characterization successfully, while imposing price-taking circumstances on Australian grapegrowers. Moreover, blending across territorial boundaries has benefited large corporations disproportionately and they don't want appellations impinging on use of a trademark, grape content, or the freedom to change wine style and grape source according to their perception of market needs. Yet in reality the Australians and New Zealanders have mimicked the territorial strategy at a national scale, by for example using Syrah and Sauvignon Blanc as their *typicity*. Ironically, corporations impose much greater control over quality and variety, than winegrowers in any territory would impose on each other.

*Ancillary and Complementary Organizations*

Winegrower associations, although the primary protagonists of a territory, work with other local or regional associations and occasionally against them. To begin with the most organizationally proximate, winegrower associations have ancillary institutions such as the French *confreries* and *academies* that reinforce the mystique of winegrowing, build a cultural linkage and foster the appreciation of wine. There are also many organizations among smaller groups within and between appellations, particularly among like-minded and equal status winegrowers helping each other out for marketing and technical matters. Unique to France, merchant associations have territorial linkages, covering a vast territory such as Bordeaux, or focusing on an area such as Libourne.

Other territorial organizations complement the winegrowers. Universities, colleges and training institutes, often at the spurring and funding of the winegrowers, help to provide human resources and general and specific knowledge. Local chambers of commerce, chambers of agriculture and other business associations work for synergies between wine and other industries. The direct and indirect paybacks are an increase in jobs, income, investment and overall vitality of the area. Of these, tourist offices often work hand-in-hand with winegrower associations. Of course local and regional governments provide organizational and fiscal support to the winegrowers and other supporting institutions. Relations between the institutions are not always harmonious. Winegrowers efforts to preserve the landscape may conflict with development plans of local governments and businesses. Even the territory, as it is carved out by a self-selecting membership, may exclude others who are represented by the regional or municipal governments within whose jurisdiction it lies. Whether in support or in conflict, however, all of these institutions are integrated into national and international systems of governance that enhance and limit territorial self-governance.

## Regulatory Governance: Empowering and Encumbering

National, state, and local government policies and laws impose governance on the wine industry in and across jurisdictions. International agreements, industry associations, and NGOs influence these laws with sufficient power to entice or impose changes on industry behaviour. The regulations may focus on governance of actual production or they may deal with health, environment, or other issues. These regulations establish a framework within which territorial organization can be initiated and operate and impose a multitude of conditions and costs. Regulatory frameworks vary from country to country, but there is an increasing convergence of mechanisms – whether formally systemized or cobbled together – that enable effective territorial organization. National appellation laws are the core of regulatory governance because they provide a territory with a defensible trademark. They are also controversial and problematic.

### *Two Territorial Frameworks*

As stated in the previous section, there are two main types of territories, those that explicitly link territorial organization and appellation and those where the appellation is supplementary but stimulates organization either by a pre-existing association or a new one. The former receive a mandate to control their reputation and the latter simply acquired a name and standard controls on production. Not surprisingly, the regulatory superstructures overseeing these territories and their appellations differ, most importantly on whether winegrowers or government bureaucrats are in charge of the regulatory process. France, which pioneered appellation systems, experimented with both.

In the first decade of the 20th century, bureaucrats initiated and administered France's appellation system. With local stakeholders input they delineated wine regions, who would be included and who would be excluded. After 20 years of frustration with the politics of this system, winegrowers convinced the government to allow them to take over administration. At the same time they brought in the requirement for *typicity* and quality control. In 1935, the winegrowers and the national government set up the *Institute National d'Appellations d'Contrôlée* (INAO) to take over the recognition of appellations, mandate authority over appellation users, and help to manage the territories.

The AOC system's objective is to provide a structure for territorial and estate differentiation. The foundation of the system is the winegrower associations. They build territorial reputations and provide a platform for estate reputations. Many associations predated the AOC system and shaped its evolution as a collection of disparate entities. The last few decades have seen a rush to organize and capture the value provided by an appellation. To apply for an appellation the associations have to generate a critical mass, define their own territorial uniqueness, and generate their own rules. The INAO assists this process, stipulating the parameters for rulemaking and assisting in geological, oenological and cultural definition, but each territory must be a collective enterprise.

The INAO, not only designates appellations, but also sees itself as safeguarding the integrity of the AOC system, preventing territory and winegrowers from damaging the reputation of AOC wines in general. It is expected to impose some hierarchical ordering of quality among winegrowers and therefore provide consumers with transparency on the many territories and the vast number of producers. To a great extent the INAO has taken this defence of vertical quality onto itself because of its long-standing belief in providing the market with legibility.[20] This role has also been imposed by external observers, who use wine territories to organize their understanding and publications and demand a coherent ordering of what has grown to hundreds of appellations. This is true of Italy and other countries as well.[21] At the outset of the new millennium, the INAO debated and eventually in 2008 imposed greater systematic controls in response to perceived threats from New World brand producers. The substance is reform of a quality control process criticized for allowing *syndicats* to conform to the minimum standard.[22] As a signal of the changes, the *syndicats* have been given the technocratic name of organizations for defense and management (*organismes de défense et de gestion* [ODG]).

Yet, despite the demand for these changes and although the formal procedures of appellation formation are often stressed,[23] there has always been substantial opportunity for organizational innovation within AOC governance. Bordeaux, and Saint-Émilion in particular, provide examples of how innovation in self-governance within the AOC system can have dramatic results. Although these initiatives require the AOC, they can chafe against the real and perceived demands of the national system.

The success of the winegrower's AOC system has been such that it has been extended to include other agricultural products, from cheese to peas. The INAO has also been entrusted with other certifying systems (Label Rouge, geographic indications, organic agriculture). A national committee dominated by winegrowers, but also composed of merchants, government representatives and other industry stakeholders decides on AOC applications, alterations to internal appellation regulations, and policy. Administration duties are the most visible part of the INAO, including support and analysis of applications (e.g. geological and historical assessments), quality control, and trademark protection.

The Italian system, although modelled after the French, constrained territorial autonomy until recently. It was initiated in the 1920s, formalized in the 1930s, revamped and expanded in the 1960s, reformed in 1992, and again in 2005. Throughout these changes and until the 2005 reforms, control over the practices allowed in specific appellations, was often decided by politicking at the national level. The structure of the appellation system was given paramount importance, but its ordering of quality was more ideal than reality. The *Comitato Nazionale per la Tutela e la Valorizzone delle Denominazioni di Origine e delle Indicazioni Geographica Tipiche dei Vini* (CNDOIGTV), the governing body of the DOC[G] system (*Denominazione di Origine Controllata [e Garantita]*) reflects external influences. It is composed of broad industry representation and winegrowers are a minority. Changes in 2005 grant wine associations (*consorzio*) greater territorial self-governance. Chianti Classico's trials over the last century testify to the implications of this slow evolution to self-governance.

Despite a long history of wine territories the US only introduced a formal appellation system in 1978. These regulations were driven by the government's need to arbitrate competing claims to geographical nomenclature and the recognition that other countries valued such a system. The US system is administered by the Treasury Department whose primary missions are to collect taxes, ensure products within its remit are marketed according to labelling and advertising regulations, and do so in a fair and efficient manner. Originally the Bureau of Alcohol, Tobacco and Firearms (ATF) designed and administered the system, but in a post-9/11 reorganization, the Alcohol and Tobacco Tax and Trade Bureau (TTB) took over. The TTB sets the standards for American Viticultural Areas (AVA) and judges the success of applications. Any interested party can petition the TTB to initiate an AVA, thereafter, others can comment on the proposal through written statements. The bureaucrats at the TTB then decide.

Although there are no prescriptions for managing the AVA after it is awarded, the process of application generates significant territorial action. Initially, AVA applications were relatively simple, but expectations for geology, climate, history, and cartography – justifications have risen. Consultants provide these credentials, requiring the sharing of costs and purposes among neighbouring winegrowers.[24] Often lawyers have to settle disputes before applications are sent to the TTB. Local cooperation could well have predated the application or resulted from it. Irrespective of whether mutual efforts are maintained after obtaining an AVA, sustained cooperation remains the volition of the community of winegrowers. Any change made to an AVA must go through another petition process at the TTB.

Despite appellation laws being designed and administered at the Federal level, winegrowers or their associations have to monitor and legally pursue any infringements of their property rights. Furthermore, the Federal appellation laws have left contestable ambiguities, particularly in relation to pre-existing use of a geographical name. As a consequence, Napa has repetitively lobbied various levels of government or relied on the courts to create *sui generis* laws to knit together an appellation system that protects its trademark. Some of these laws protect only Napa, others offer protection to other territories. Ironically, in 2008 when the Italian authorities suspected some Brunello de Montalcino winegrowers of using non-appellation grapes and varieties, the US TTB inspectors shut down imports, even though the US allows producers to use some European appellations (e.g. burgundy and champagne) as "semi-generic" wine names.

A US-styled system, focusing on geographical delimitation and grape content, is used in Australia (Geographical Indications, GIs) and New Zealand and Northern Europe as well.[25] The other European countries and Latin America (Spain's *denominación de origen*, DO; Argentina and Chile's *denominación de origen controlada*) and Canada's VQA follow the French model. Thus in the three territories of this study we have reasonable examples of the spectrum of appellation styles. All appellations, however, are not territories.

In all countries blended wines are classified by regional and political boundaries, such as county, multiple counties, other political boundaries or a traditionally defined region. Although stakeholder input is used to make up these appellations, it is more of bureaucratic exercise and the boundaries are usually set according to political borders. These appellations are usually large and allow blending from diverse geographical conditions. In France and Italy, because of the assumption that *terroir* based wines are favoured with better biogeoclimatic conditions and greater intensity of human care, the AOC wines are formally set on top of a hierarchy. The wines at the base, geographically and qualitatively, allow greater spatial blending of wines. The broader appellations allow practices prohibited in AOCs and conversely deny some AOC practices (e.g. in France table wine, with no spatial limitations, is not allowed to put a vintage on the bottle).

The US has a similar spatial structure that allows blended wines some semblance of connection to location. Horizontal quality differences are implied in the US labelling framework, but there are no official vertical quality differences. The US also limits the use of estate bottling to wineries in AVAs. In contrast you can estate bottle in a French or Italian regional classification. Broader appellations are not without collective action, but it does not reach the intensity or complexity of territorial appellations. The greater degree of blending, fewer restrictions on the import of grapes and the location of vinification, traditions of low quality and bulk sales, and simply a greater space, hinder collective action. That does not mean that territories cannot be carved out of these areas. Increasingly they are, as winegrowers, old and new to the area, realize the benefits of differentiation and collective action.

*International Recognition of Appellations*

An appellation's stimulus for collective action would be of little use if the trademark is not protected nationally and internationally. That recognition comes in a range of regulatory and non-regulatory mechanisms that vary from country-to-country. The regulations are evolving in a struggle to create space for territorial intellectual property rights in a global system favouring corporate intellectual property rights.

The legal protection of territorial rights dates to the 1700s with the delimitation of Portugal's Douro valley. Since, sporadic attempts have been made to create guarantees of regional authenticity, within and outside of Europe. The Paris Convention (1883) was the first international treaty to include indications of source and appellations of origin among industrial property rights. This agreement was revised several times through the 20th century and was signed by 160 nations as of 2000. The Madrid Protocol (1891) and the Lisbon Agreement (1958) increased protection for intellectual property derived from origins, but these agreements garnered fewer signatories. More broadly, respect for geographic origins is recognized in individual countries with a variety of trademark, business practice laws, and *sui generis* or special laws.[26] Undoubtedly, however, the evolution of France's AOC system stimulated France's recognition of territorial property rights, and in turn, spurred adoption by the EU. The EU then led an international drive to respect appellations of origin (AOs) and geographical indications (GIs).

Thus far the propagation of territorial property rights has culminated in the TRIPS[27] agreement that binds all WTO members to respect AOs and GIs and to provide for minimum standards of enforcement and a strong dispute resolution mechanism.[28] To accomplish this goal, the WTO had to deal with different property rights systems among member nations and non-members. In particular, the Europeans' continental system provides for the state to proactively protect territorial rights, while the Anglo-American tradition leaves the onus on the damaged party to initiate legal action.[29] And with TRIPS enforcement is left to domestic mechanisms. Such differences are consequential for the cost of territorial defence. Other problems remain. Some non-EU countries allow the use of European place names to designate a style or brand, so-called generic names, which have little relation to the original. The EU used the size of its market to compel Australia and South Africa to forgo this use, but has had less success with the US. Furthermore, although TRIPS doesn't allow proprietary trademarks to include geographical indications, eliminating pre-existing uses or precluding new uses of these names requires court battles in the US, of which Napa has much experience.

Other issues complicate the intellectual property claims of territories. So-called "traditional expressions" and oenological practices have become bundled with different regional, national or continental systems. For example, the winegrowers of Chianti Classico fear the term "Riserva," which they use to designate aged and higher alcohol wine will be devalued by indiscriminate use. Similarly, the EU at one time proscribed using oak chips to flavour to wine, among its winegrowers and in imports, to prevent the debasement of the use of barrels in appellations. The World Wine Trade Group formed in 1998 to eliminate these and other perceived barriers. The New World countries of the US, Australia, Canada, Chile, South Africa, Argentina and Brazil signed the Mutual Acceptance Agreement on Oenological Practices to allow exports to each other irrespective of differences in winemaking practices. The organization is a counterforce to the EU's bilateral agreements that expanded GI property rights and traditional expressions, and the Paris based OIV (*Organisation Internationale de la Vigne et du Vin*) whose mission to harmonize internalize standards was considered too Eurocentric. The battle for territorial property rights involves other players. For example the Grocery Manufacturers of America, representing over US$ 400 billion in yearly sales, fought against any expansion of protection of geographical indications for food and beverages, claiming (without a hint of irony) damage to long-standing trademarks and brands.[30]

The stances of countries on the international stage, of course, reflect internal debates and institutions. In Europe the battle for AOs and GIs, despite occasional North European indignation, has been won and strong laws and institutions protect them. Not incidentally, these laws reduce the costs of winegrowers in defending their intellectual property. Government fraud investigators in France and Italy constantly monitor the veracity of wines in bottles with appellation labels and are important parts of the appellation systems. The INAO is particularly active in defending use of AOC names in France and around the world.

*An Efflorescence of Collateral Regulations*

Although appellation systems vary by country, insofar as they provide trademark protection and stimulate collective action, they provide an associational platform to deal with other regulations that provide direct and indirect challenges to winegrowers. These regulatory issues include agriculture, health, labour, pensions, alcohol taxes in domestic and foreign markets, distribution barriers, and the environment. Winegrowers and territories, because of scale limitations, also look to government for investments in R&D and education to enable them to create the differentiation tools and to compete globally. To achieve these ends, governments are lobbied at local, regional, national, EU and global levels. The production, transaction and opportunity costs of lobbying for and against these many policies far outweigh the costs of winegrower associations and the running of an appellation. Indeed lobbying is among the most important functions of winegrower associations. Lobbying usually requires federation of the association into a greater force as in done in Europe, or the fractious alignments between groups such as the Napa Valley Vintners, the Napa Valley Grapegrowers, the California Wine Institute, and the Family Winemakers of California.

Describing particular regulatory challenges facing each territory will be left to later chapters, but a few illustrations integral to those discussions are offered here.

Europe's Common Agricultural Policy (CAP) is infamous for both incentives that produced its wine lake and its struggles with national governments to devise means to reduce the surplus. Control over planting rights is one mechanism for performing the later. Countries are given a quota, interrelated with incentives to grub up vines, and each devises different mechanisms for domestic allocation. Winegrower associations lobby for their interests and in some cases are entrusted with devising means of allotment of planting rights and grubbing up. At the individual level, while some winegrowers look for greater assistance, others chafe that expanding their vineyards depends on local, national and continental politics. Indeed as CAP policies are subject to WTO scrutiny, the vineyards are controlled at a global level. Americans are also influenced by agricultural policies such as state laws and tax exemptions that encourage wineries[31] or dispense irrigation water at below market rates.

Health is a huge issue for winegrowers. They must invest in all the equipment, labour and paperwork needed to comply with the extensive regulations concerning plant health and chemical residues, sanitary conditions in the winery, chemical additives to the wine, and labelling requirements related to inputs (such as sulphites). Winegrower associations clarify these regulations and keep winegrowers updated on changes. State or National regulations are influenced by the UN's Codex Alimentarius Commission standardization of food code regulations and the WTO's efforts to ensure that national regulations don't become technical barriers to trade. Wine's alcohol content carries the health issue into related realms. US prohibition lives on in the relegation of wine governance to the TTB and the control given states over the distribution of alcohol. The later resulting in a wholesaler monopoly situation, particularly vexing for direct sales of estate wines, but a spur to winegrower association. Driven by powerful NGOs, neo-prohibitionism in the US is a constant worry to the industry and its lobbyists. The anti-alcohol movement is also powerful in France and supported by the health ministry. Laws severely limit marketing efforts, and few issues rankle the struggling *vigneron* more than fighting her government to provide what she believes is part of a healthy lifestyle.

Environmental governance is both challenge and opportunity for winegrowers. Local and regional governments control winery effluents but also subsidize the equipment to bring it under control. Pesticide and fertilizer use are scrutinized by regulators, NGOs, local communities and customers, but their reduction also lowers costs, responds to demands for more natural products, and reinforces *terroir* differentiation. Biodiversity and protection of habitats may require vineyard and income to be given up, but supports the collective effort to attract wine tourism. Most territories now prioritize environmental protection, advising on vineyard and winery practices and initiating territorial approaches. The latter tied to reducing encroaching urbanization and finding a means to live with the non-winegrowing community within the territory.

## Distribution Chain Governance

The term commodity chain reveals a prejudice towards large quantity production of homogenous goods, and at relatively low value. Conversely, the term supply chain depicts a bias toward assembly or distribution firms controlling inputs from a lengthy linkage of suppliers. The value chain[32] is less prejudiced, pointing to the potential for value to be created anywhere from initiation of production to consumption. There have been several adaptations of this concept, notably the distinction between producer and buyer value chains that emphasize a shift of control from final-product manufacturers to distributors.[33] In the wine industry, value can be generated anywhere from the vineyard to final appreciation, but while keeping the value chain as reference, I use a bias that originates with the winegrower's need to maintain control over differentiation. The winegrower's perspective is for control of a distribution chain. This breaks the convention in both producer and buyer value chains of value-added being concentrated near the market. The winegrower wants to claim the value normally captured at the market end of the value chain by distributors or integrated producers.

The distribution chain is governed by the transaction costs of bilateral relations between firms[34] in relation to the conventions among firms in the industry (or *filière*) and the many formal institutions whose role it is to mediate among transaction partners.[35] The nature of this governance changes according to locations in the global production network.[36] The complexity of the supply chain is one of the reasons to integrate it in corporate form. Integration eliminates many uncertainties and costs between supplier and buyer, brings more power to institutional governance, increases freedom from conventions, and enables smoother exports, imports and distribution. Correspondingly, the winegrower needs to develop some mastery over their distribution chain. Rarely can they do this alone.

*The Wine Value Chain*

The value chain of the wine industry combines the main stages of grapegrowing, winemaking, blending and wholesaling, retail, and consumption. With the exception of consumption these stages can be integrated completely by estate wineries or large producer-distributors. Alternatively, a few stages can be integrated as is done by merchant-blenders, cooperatives that do their own marketing or retail firms that organize production. Or, independent firms may operate each stage, the linkages mediated by market-based transactions, with or without coordination for a distinct product.

The value chain can be extended backwards into land ownership and forwards into recycling. The former is becoming increasingly important as winegrowers look to increase not only the value of their wine but also their land. Many winegrowers in Napa, Chianti Classico and Bordeaux claimed that raising property value by building a wine's reputation is a surer way to make money than selling the wine. Financiers have noticed this escalation and turned vineyards into investment vehicles. Recycling and other ways of reducing environmental impacts farther downstream in the value chain are becoming important as the industry attempts to reduce its ecological footprint. Winegrowers, their distributors, and their institutions must respond to environmental governance in their markets and in some cases are changing their packaging to do so. Parallel to the main value chain is provision of services and capital equipment, including: vineyard management; viticultural, winemaking, and marketing consultation; "custom crush" (rented winemaking) facilities; labour contractors; barrel makers; winemaking equipment; educational and training institutions; and so on. Moreover, the media, wine specialists and others depend on and foster this value system.

The products within the core value chain include grapes and must sold by grapegrowers to estates, small or large wineries or to cooperatives. Bulk wine is sold by winegrowers to the same outlets, cooperatives excepted. Winegrowers and wineries can sell directly to consumers or to outlets such as specialist shops, restaurants, merchants (as their own brand or as the merchant's brand), wholesalers, and supermarkets. Merchant-blenders can sell their own brands onto wholesalers and retailers or they can bottle wine for retailer sales as "own brands." Merchants also sell estate and other brands on to consumers, restaurants, wine shops and supermarkets.

## Transaction Types and Tensions

Transaction modes include spot prices for grapes, must or wine; brokered sales; contracts binding supplier and buyer over time and imposing quality parameters; futures markets for unbottled wine; direct sales; and wine clubs. Inherent to all these modes are tensions between seller and buyer and the attempt by winegrowers to differentiate.

The use of spot markets remains an important feature of wine regions. It provides base prices for commodity and differentiated grapes and wines through highly competitive markets. Prices for grapes may vary significantly within the short harvest period and grapegrowers must quickly find buyers for their highly perishable produce. Conversion into bulk wine provides protection against this pressure, but only a few months of relief, and prices can vary greatly as regional or worldwide supply rises and falls. Actual transactions occur through repeat exchanges or through various forms of contacts, but to reduce transaction costs, sellers and buyers usually use brokers, occasionally using electronic auctions.

Spot markets have two outstanding problems. First, information asymmetries can exist between grapegrowers or winegrowers and vintners and/or merchants. The growers don't know the prices paid in other transactions or of the total supply and their bargaining power is therefore low. Buyers gain information through negotiation with many producers and brokers in or outside of a region, but it is more difficult for growers to gather such information because its release will be resisted by both buyers and competitors in order to preserve their competitive positions. Second, spot prices commodify products within a limited range of characteristics.

Quality can still be important, but as the wine is destined for blending there will be little return for *terroir* or other differentiation. Indeed as merchant-blenders are interested in producing their own style and brands, it is against their interest to recognize differentiation efforts unless they want to use the *terroir* effect. When they do, they will likely have to pay for it.

Beyond spot prices, virtually all wine regions exhibit more complex bilateral relations between different stages, between grapegrower and winery, winegrower and merchant-blender, grapegrower and cooperative, an estate and merchant or even between estate and retailer. Within any of these relations there is the potential for self-interested strategizing where exchange partners can be competitors or collaborators, or both.[37] Whether, mediated by bilateral negotiations, brokers, contracts, or organizational confrontation, the sell-buy relationship remains inherently problematic because of the necessity to coordinate each party's roles while negotiating over shares of costs and profits. The strategizing and attempts to capture value can lead to under-investments in viticulture, viniculture, and marketing. Variation in the weather change quality and quantity from year to year to make these relations tenuous, but most important is the highly elastic nature of agricultural prices. Prices can rise rapidly if there is a modest scarcity of supply and can crash if there is a modest surplus. Irrespective of the potential upside, the reality for winegrowers is that despite growth in consumption there is still a surplus of grapes; sellers vastly outnumber buyers and can go elsewhere for grapes or wine. For both sides in these vertical relations there are significant tensions to overcome, and they and their territory will lose out over the long term if a mutual solution can't be found.

The governance of bilateral relations is not simply a matter for transaction partners, any inefficiency in their bargaining, irrespective of long-term benefits, will not only be penalized by market forces, but invite regulation. If employment or other welfare conditions degrade, then government may step in to impose settlement. On the other hand large firms may exploit the tendency within the value chain to impose double or multiple margins on consumers. Regulators may well applaud the integration of the supply chain for enhancing consumer welfare.

A focus on bilateral relations and transaction cost governance[38] marks much of the analysis of vertical relations in wine regions. That focus overshadows a related and more important requirement for estate winegrowers.[39] Although it may seem counter-intuitive that a small producer will need a large spatial market, differentiation often means that consumers for their product are widely distributed. A distinctive producer often needs a global market to take up all their production. To access their widely dispersed customers small producers need many distribution channels, and for consumers to benefit from the variety produced by estate winegrowers, there must be sufficient intermediary distribution channels. This chasm can be bridged by direct sales, whether door-to-door, winery door, or Internet sales, but these approaches have limitations. Intermediaries, such as wholesalers, merchants, and retailers provide access to many potential customers while providing customers with information, compatible products and other services and reduce regulatory, storage, marketing and distribution costs for the producer.[40] However, it is risky to rely on exclusivity agreements with merchants in any market to avoid distributor complacency. Another obstacle to multi-channel efficiency is that increasing differentiation decreases the capacities and incentives for merchants to represent dozens of chateaux adequately. Merchants are compelled to free ride on reputation of chateau and appellation,[41] and thus the onus returns back to territorial organization.

*Territorial Resolution*

Territorial winegrower associations are the main institutions governing the distribution chains for independent winegrowers. In France and Italy, they are federated into national organizations, and as they receive their mandate from the government, there is significant integration with regulatory governance. In the US there is more of an ad hoc grouping of organizations at different levels. In all countries, associations can be distinguished by those focused on general industry needs and are inclusive, and those oriented towards subsector needs, such as *terroir* winegrowers, family winemakers, merchants or large firms. I leave the extended description of the actual institutions, their activities and relations with other institutions until later chapters, outlining their mechanisms of distribution chain governance here.

Bringing transparency to the information asymmetries inherent to the spot market is a common thrust. Transparent pricing is a reference point for both sellers and buyers and the source of analysis for various services offered to them, such as brokerages. Typically, government support or legislation is required for the gathering and dissemination of sales information. The enactment of such legislation, however, results from the collective action of winegrowers, such as the Bordeaux Wine Board linkage of winegrowers and merchants. Such organizations are typical in French wine regions. Giraud-Héraud et al.[42] claim that they ameliorate information asymmetries and help to develop stable contractual arrangements. Similar needs and actions resulted in California's crush report and the setting of a floor price for Chianti Classico grapes. This commodity prices transparency remains important, but is not sufficient for the proliferation of estate wineries. Winegrower associations increasingly seek to clarify the relationship between buyers and sellers of more differentiated products.

Contracts are a powerful means of stabilizing the relationships between sellers and buyers. They may involve a mix of many variables, such as quantity, quality, delivery times, information sharing, vineyard designate rights, price setting agreements, and penalties for not meeting such conditions. These contractual arrangements rarely lack influence from the norms and innovations of the territory. Moreover, the innovations are pushed not only by market demands, but also by the political-economic power of the two sides as grape and winegrowers organized themselves to obtain better contractual arrangements from wineries and merchants. Napa has experienced this contention as a confrontation between two opposing organizations, but the Bordeaux Wine Board attempts to institutionalize negotiations between buyers and sellers. The imposition of quality controls by territorial associations is also a mechanism to preclude and reduce the transaction costs of these contracts, and for the winegrower increases potential buyers.

Spot markets favour large firms because competition among many supplying firms with no market power allows them to purchase the raw ingredients for standardized production at the lowest price. Inversely, estate winegrowers strive to create a market for differentiated production, a system where buyers will be attracted to and possibly bid on unique products. Bordeaux's *en primeur* sales is the world's most powerful example of a differentiated futures market for agricultural products, but other territories are trying to emulate its success. In Bordeaux's case *la place* is a key platform to maintain the linkages between the myriad chateaux and hundreds of merchants they depend on. Other markets include trade fairs organized at territorial, national and global levels or wine fairs that bring differentiation into the homogenizing tendencies of supermarket oligopoly, internet portals, territorial wine shops, and charity auctions. Creating these markets requires collective organization and often the entertainment of thousands.

If territorial reputation is important, then intuitively it makes sense to expand reputation by advertising and there is a high level of support for generic advertising, including making it compulsory for producers in an industry or region.[43] Bordeaux has followed the generic strategy to service the proportion of its production sold in relatively large volumes. Such promotion was also thought helpful to the estates. But as territories try to reduce tension in the distribution chain they must be wary of tension between collective action and the interests of individual winegrowers. Connecting the benefits of generic promotion with individual returns is difficult, particularly when individual producers are trying to establish their own brands. It has been found, for example, that generic advertising can hurt differentiation efforts[44] and that while wine regions can potentially benefit from combined advertising campaigns, their differences may also undermine unified messages and impact other collective efforts.[45] Marketing can be particularly divisive because it is the most costly of all collective actions. It is not, however, the only collective activity with contradictions. R&D requires economies of scale, that is contributions from all producers to be viable, but again, it is difficult to tie R&D results to each producer's profit potential.

The tension between territorial and estate marketing is heightened by the lure of brands. Corporate trademarks, privately owned and transferable, can be authoritatively protected and supported. On the other hand, the appellation is not privately transferable. Because of this difference, and also because of the volume of marketing literature extolling brands, some winegrowers put their faith in a brand that, in addition to selling, can be sold. Selling a brand is generally not a problem unless, the buyer's intention conflicts with the objectives of the territory. The inability to transfer the regional trademark is partially compensated for by the transferability of land. In regard to these tensions, marketing and other collective efforts have to be devised to respect and support the estates while building the territory.

A powerful tool to bind winegrowers together and tie merchants to territory, is to offer what the corporations can't – the images and values produced by the territory as a whole. The images include the landscape and environment, the culture created, and the aura of a territory of family wine estates. A company can be a part of these, mimic or portray their image, but it cannot generate all the complexity and diversity that a wine region offers. However, maintaining the landscape and culture of a region is not without financial or participatory costs, nor does it occur without organization. Thus the cost of generating these positive externalities must be internalized and shared by all members of a region. This internalization can be expensive, especially when vineyard or winery practices, that damage the environment have to be curtailed.

The responsibilities and benefits of social reproduction is another area where territorial organization differs from the corporation. Health care and pensions, as examples of social costs, are occasionally internalized through exemplar corporations, but for large segments of the labour force, they are not. Broader social institutions within a territory's nation or region, therefore become important foundations of support for the industry. Where they are lacking winegrower organization needs to provide alternatives or risk sullying the reputation of their region. The sharing of externalities brings us a final issue determining the viability of the distribution chain, the discount rate held by winegrowers. With commitment to a territory there is greater likelihood that the resource, in this case the reputation, will be respected and maintained.[46] Conversely if people expect to make fast returns on investments, and to disinvest, then there is greater likelihood that the reputation could be exploited without commensurate investments and the sustainability of value chain compromised.

## Summary

Estate winegrowers are inherently tied to the fortunes of their territories. Even the most renowned wines benefit from being the best from their territory. Thus for the winegrower, although they are primarily interested in the differentiation of their own product, they need to utilize the reputation of their territory. To do so they have to help build and maintain that reputation, take part in collective governance and follow rules. Depending on the national context, self-governance is built with varying regulatory support, and in any case territories lobby to get regulations on their side. The winegrowers' ability to influence the distribution chain, similarly, depends on the effectiveness of collective action supported by government or without. But influencing the distribution chain is not the only use for the territory as collective. The direct costs of running an estate, sanitary and environmental regulations; labour, health and pensions; profit, alcohol and other taxes; plus all the other regulations governing advertising, planting, etc. – all these absorb enormous efforts to understand and deal with.

The core demand for self-governance of a wine territory is to balance needs for both differentiation and collectivity. Depending on the issue, effective incentives, rules, monitoring, and enforcement systems and collection of dues have to be imposed to support both territory and the ambitions of winegrowers within it. This is problematic as there is ample room for free-riding and back-scratching detrimental to territorial reputation and for the reputation to be influential it has to be built-up and maintained over the long term.[47] The need for effective self-governance of the territory, for influence over the distribution chain and regulatory environment is being pressed by global competition. Thus although wine territories may have evolved over centuries, each with a fascinating history, the interrelations between land, people and their wines need to be analysed for their organizational effectiveness. That is the approach taken in the next chapters.

# Chapter 3
# Bordeaux:
# From One to Ten Thousand Chateaux

London 1660. A man walks into a pub and says: "can you pour me a Ho Bryan?"[1] Thus was the estate winery introduced, identifying a wine grown on a particular piece of land, and marketed under that brand. By 1665 the Pontac family, proprietors of the Haut-Brion estate, initiated direct sales by opening and operating their own tavern. They even differentiated their brands into first and second wines, giving each a separate identity. Back on their estate, they made investments in improved wine equipment, techniques and aged the wine. Haut-Brion was followed by new estates established in the Médoc – Latour, Margaux, Lafite – that would also sell their particular wine under their own names. These proprietors broke the merchant's practice of blending all wines into a homogenous product, developed market power for themselves and added value to their products. The essential innovation was the concept of the estate winery, the linkage of both quality and differentiation to characteristics of the plot of land from which of wine originates. Originally, this linkage was conceptualized in the use of the term "cru" (growths) to describe plots of land that would consistently turn out great wine. Later, as winegrowers took more control over their marketing they used the title "chateau" to give distinction to their property and their brands.

The concepts of the cru and chateau would diffuse throughout Bordeaux over the next 200 years and today the majority of its 10,000 winegrowers bottle a least a portion of their production under their own label. Indeed, the name Bordeaux, not only identifies the origin of many of the world's finest wines and their estate basis. As significantly, Bordeaux as a territory created a platform for the generation of variety on an unprecedented scale. Consider, for example, that the number of Bordeaux winegrowers easily outstrips the number of all the wineries in the Argentina, Chile, South Africa, the US, Canada and Australia combined. Creating this platform for variety, however, has been a long evolution, marked by several interdependent challenges.

Haut-Brion was the exception to a rule that was as powerful in the 18th Century as it is today.

> … the creation of a cru (an estate wine) involves landownership, the making of the wine and its marketing, and these are three activities which it has never been easy to integrate. The more numerous the crus and the market in which they were sold, the more difficult it became to combine all three elements.

Inevitably, the majority of owners have always had to rely on the merchants ...[2]

The merchants integrated the region by sourcing for their markets, but within what started as a commodity trade, the winegrowers developed their *terroir* and brands and other territorial reputations. Merchant dominance gave way to a more balanced relationship, but one that has always needed mediation to some extent. Governance has been most obvious in the mediation of bulk transactions between winegrower and merchant, but more challenged to provide both coherence and opportunities for differentiation to 57 appellations and 12,000 chateaux. That governance comes from a large number of formal and informal institutions, operating at different levels. This chapter looks at the institutions integrating the entirety of Bordeaux and in the next chapter we turn to how reputation is created at the territorial level.

### Integrating Variety

Territories often derive cohesiveness from a shared landscape and climate. Bordeaux's expanse is tied together by the Dordogne and Garonne rivers where their confluence creates the Gironde estuary. The territory extends for 50 or more kilometres along each of these three waterways and reaches inland from their banks for up to 30 kilometres. The limestone gravel of the southwest shores of the Garonne and Gironde and the limestone escarpments of the northeast banks of the Dordogne and Gironde are thought to unite the region geologically. A regional climate, moderated by the Gulf Stream and sheltered from Atlantic winds by a pine forest to the southwest adds further cohesiveness. Together, this is the world's largest fine wine region (Figure 3.1), consisting of 113,384 hectares of vineyards, extending 105km north-south and 130km east-west, and covering most of the *department* of the Gironde. This landscape is fragmented by meso and microclimates, and a mosaic of geologic, soil, relief and aspect differences. These differences exist between the Médoc, Entre-Deux-Mers or Blaye, and they exist within any of these areas. Bordeaux's territorial identity depends more on an integration of variety than in homogeneity.

Bordeaux's integration as a region originates in the trade that occurs within it. Merchants collect and blend bulk wine from around the region. Brokers (courtiers) assist the selection of wine by matching buyer and seller and guaranteeing the terms of the contract, gaining 2 percent for their efforts. In the past the merchant-blenders (*négociants-éléveurs*) brought the wine to their cellars in Bordeaux for ageing, blending and exporting, but now these facilities are set within bulk producing areas such as Entre-Deux-Mers. The merchants also integrate by assembling portfolios of different chateaux, appellations, or styles of wine for combined global distribution. The merchants are a diverse lot, some blending, aging, bottling, and distributing, others performing only one or two of those steps. Many own chateaux, operate on leasehold and manage others, often using that estate's own name as one of their chateau labels. Some specialize in Grand Cru or

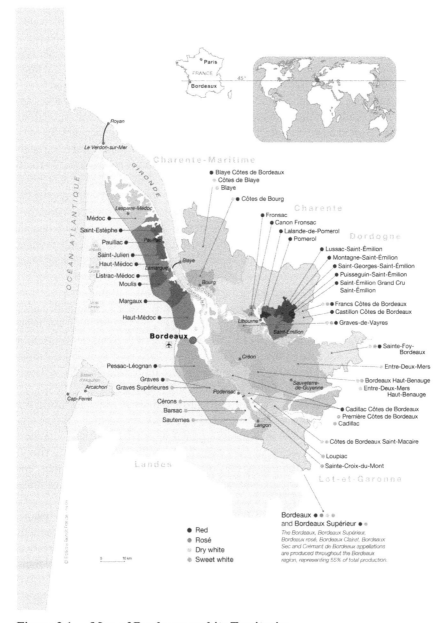

**Figure 3.1     Map of Bordeaux and its Territories**

specific appellations, others on brands, bulk transaction, a selection of chateaux, or biodynamics. Merchants may have to compete for a share of a chateau's production, hold exclusive distribution contracts, or be a nominal representative. Supermarket, café, hotel, restaurant, individual consumers, export destinations and other distribution outlets may be niches or provide economies of scope.[3]

Most merchants continue to be based in the city of Bordeaux, but Fronsac, Pomerol and Saint-Émilion are also served by *négociants* operating out of Libourne. The merchants number about 400, but just 25 of those account for over 80 percent of turnover. Several merchant houses have traditions traceable through hundreds of years. A few, such as Barton & Guestier or Ginestet, have become brands that have been bought and sold, while others like Calvet, Schroder & Schyler or Sichel remain within the ownership of the same lineages (if not necessarily under their management). A hybrid of those two approaches, C.V.B.G.-Dourthe-Kressmann, with brands such as Dourthe: Numéro 1, Kressmann Monopole and Delor Reserve, claims to be largest seller of Bordeaux wines. More recent entrants have also made large impressions. The Castel Groupe was started by nine siblings in Bordeaux and has become France's largest wine distributor and one of the top 10 in the world. Most entrants have not grown as prodigiously, but fill niches such as restaurant orders, direct chateau to home sales, sales by expatriates to their home countries and so on. The expansion from winegrowing into distribution has been natural for many estates marketing their own wines. Baron Philippe de Rothschild broke this ground, not only on marketing his chateau, but also in building the most famous and highest volume brand and representing and managing other estates. Many other winegrowers have followed in the last few decades.

French multinationals such as Pernod-Ricard maintain a presence, while the Americans, British and Japanese have made occasional forays. Many merchants have significant sales of different French regional or other countries' wine and even some winegrower-merchants work with wines from outside the territory. However, the most notable characteristics of Bordeaux's merchants are that, not only do they originate in Bordeaux, but the great majority of their sales concentrate on Bordeaux wines.

The merchants are distributing the production from about 10,000 separate landholdings, the vast majority of which are independently owned and operated. The properties can be divided into bulk or generic producers, cooperative members, the *petits chateaux*, and the Grand Cru. About one quarter sell all their production as bulk wine to the merchants. Many have entered into comprehensive management or quality control agreements, and depending on the conditions, the merchant may use one of these holdings as a chateau label. Another one quarter of the proprietors, by landholdings and by volume, send all their grapes to cooperatives. Much of the cooperatives production is marketed directly and some sold to merchants. The winegrowers are compensated by a portion of sales. Legally they can sell their production as estate bottled, blending from vineyards within a territory, and selling as the cooperative's "chateau" label. Increasingly, to meet the demands of an estate driven market the cooperatives are selecting and bottling wine from individual estates or from specific vineyards. Together the cooperatives and independent producers' contract wine that the merchants sell as chateau bottled accounts for 13 percent of total production.

Chateaux bottling was undertaken by a small minority up until the 1970s, but by 1980 about one third of Bordeaux could claim to be *mis en bouteille à la propriété*. In the subsequent decade the proportion climbed to half and has stayed at about 50 percent since then. All told, there are about 12,000 wines carrying different labels. The great majority of these 5,000 estate winegrowers are known as *petits chateaux*. They bottle at least some of their own wine for direct sales to private customers, to retail outlets, and most likely to merchants. They will also sell a varying amount, usually their second or third quality wines as bulk to merchant-blenders. These firms vary from those making excellent wines to those making wines indistinguishable from or less than favourably compared to generic wines. The Grand Cru, strictly speaking, are about 100 chateaux that have achieved the distinction of classification. Depending on the classification, the distinction may be accorded based on *terroir*, market evaluation, tasting of the wines, marketing activities, and other lesser criteria. Less strictly speaking, other chateaux have achieved consistently high evaluations by market and critics, and can be thought to have entered into a rarified critical and market-based evaluation. Perhaps the most reliable distinction between a modern Grand Cru and the *petits Chateaux* is acceptance into the futures market (*en primeur*). The Grand Cru will also sell part of their production as bulk, but they have more incentive and success as selling it as second or third wines (brands).

Differences among the chateaux accrue from size of holdings, capitalization, winemaking approaches and creativity, marketing and so on. Particularly at the top end, insurance and luxury goods companies, retired tycoons and others with deep pockets and who are not necessarily concerned about profitability have bought trophy chateaux. But people searching for a new beginning have brought new blood into all corners of the region. Some owners hold many properties, seeking managerial and other efficiencies. Most chateaux are single entities and family operated. These differences among the chateaux are significant, but many derive from and are overshadowed by differences based on territory and *terroir*.

Bordeaux's physical and production geography is divided up into territories. The territories usually have a landscape endowed with the proclivity for making great wine, but they are not limited to those confines. For the most part the boundaries coincide with political demarcations at the scale of a commune or canton, usually combining several in whole or in part. Although maps of the territories usually show them covering the entire surface area within their boundaries, land that can produce AOC wine is a smaller portion, surveyed to a vineyard level. Winegrower associations, *syndicats*, define the 35 territories, with each using one or a few of Bordeaux's 57 appellations. To use an appellation allowed in a territory a winegrower must follow the appellation's regulations on grape varieties, viticulture and winemaking. The largest territory, Bordeaux and Bordeaux Superieur, encompasses the entire Gironde and its *syndicat* governs the so-called "generic" Bordeaux wines. These can be white and red, superieur, rosé, clairet, crémant but can be blended from across the whole territory or parts of it. The 35 territories are also grouped into families based on proximity (e.g. the Médoc and Graves or Libournais). Some types of wine (e.g. the white wines) have their own *syndicats*.

Territorialization has consequences. For example, in the communal territories/ appellations of the Médoc, 94 percent of wine is sold under winegrowers' brands. In Entre-Deux-Mers, although 61 percent of winegrowers have brands, they sell 59 percent of wine in bulk. In the broader Médoc, cooperatives account for 11 percent of production, but in a few territories, they produce close to half. Average prices paid for bulk and bottled wine, and per hectare of property vary by multiples among the territories. These averages break down within territories where *terroir* creates distinctions recognized by classification and markets. *Terroir*-based distinctions, by-and-large show remarkable continuity, even after accounting for consolidation of holdings, and changes in ownership and relative performance. The rise of new reputations, from properties converted to bottling from bulk, and particularly in generic dominated territories, is not however precluded. While many of these within-territory distinctions may seem problematic, they are generated or legitimized by the territories. At the same time the chateaux accept the governance of the territory over their winegrowing activities. The territories integrate variety for collective objectives. A similar form of governance is used for Bordeaux as a whole, because although trade draws it together, inherent conflicts need to be dealt with.

### Trade, Differentiation, and Governance

Although the Romans founded winegrowing and trade in the region of Burdigala, the territory as we know it today has its roots in exports to 12th century Britain and evolved with significant public and private management of its fortunes. When Aquitaine and Bordeaux became part of England with Alienor's marriage to Henry II, tax exemptions and royal purchases secured Bordeaux a market. Yet, the quality and quantity of local winegrowing didn't suffice, necessitating supply from the high country, up the Tarn river to Galliac and the Dordogne to Cahors. To develop local vineyards, Bordeaux's merchants, proprietors, and burghers obtained the *privalège de Bordeaux* in 1214 to force[4] the high-country wines to wait until late November before exporting through Bordeaux. In the 1400s Médoc wines similarly were compelled on a upstream detour to Bordeaux before export.[5] Until the 16th century, the merchants, the proprietors, and parlementarians protected each other's interest. The relationship changed when foreign traders arrived with their own tastes and the ability to modify the wine according to those desires.

The Dutch were the first to plant varieties and blend Bordeaux's wines for their own tastes and others adopted the practice. The "*travail à l'anglaise*" blended for a more homogenous product and changed the style of wine. Bordeaux's relatively light wines, even Médoc growths, were fortified with wines from the *palus* of Bordeaux, the high country, the Midi and Spain.[6] Some were re-fermented, but most were blended, racked and aged by the *négociant-éléveur* (merchant-blender) in their cellars in Bordeaux or in London. The most famous merchants were known as the *Chartronnais* after the quai in Bordeaux where they congregated their cellars. In the 1700s, traders, such as Johnston, Lawton, and Barton, exported to Britain to capitalize on the aristocracy's interest in the grand cru. Later, Germans, Swiss, Danes and other Europeans imported generic Bordeaux to their countries, and captured a share of British cru exports. Much has been written of these "wine masters,"[7] their impact on Bordeaux's grand cru and their legacy in many brand names. There was also hundreds of other *négociants* who, over the centuries, exported Bordeaux to distant corners of the globe, adroitly sustaining and creating markets through the vicissitudes of war and trade regimes.

Despite the distribution gained from the merchants, the winegrowers were aware that they were disadvantaged in the relationship. Following Pontac's lead, they responded with a strategy that, while complementing the *négociant's* trade, also challenged their control. The winegrowers separated property into estates, distinct parcels, chose grape varieties, graded their wines, and created distinct styles. Even the now famous Bordeaux blend – cabernet, malbec, and merlot – was part of the winegrowers response to *négociant* knowledge of market demands and to the full-bodied wine produced by the "*travail à l'anglaise.*" The merchants were not opposed to differentiation per se, indeed, the Chartronnais learned to benefit from differentiation. Discerning quality differences was important not only to horizontal differentiation, but also vertical. Since the early 1600s the merchants and proprietors, with the intermediation of brokers (*courtiers*) negotiated base prices for quality classes on a yearly basis in order to improve their blends.

Yet, the classification system also reveals the inherent tensions of the négociant-vigneron value chain. Brokers reduced transaction costs by visiting all the properties and assessing the wines, and over time the merchants sought to further minimize these costs by avoiding yearly adjustments to the classification. The institutionalization of a hierarchic classification gave the proprietors price stability, but they lost the revaluations that reflected changes in quality. For example, when the Bordeaux Chamber of commerce took this ranking to the Paris exhibition of 1855, all sides realized the attention it would attract, but the merchants didn't want to disrupt the existing ranking with a reassessment.[8] They were still primarily interested in blending the vast majority and even the highest ranked wines.[9]

Ten-year buying contracts, *abonnements*, were another innovation within the value chain to share risks, but unfortunately they were still susceptible to speculation from both sides and to wild price fluctuations in the markets. *Négociants* captured the lion's share of increases in value while the winegrower's incentive to maintain quality declined. The *négociants* also felt the income variations resulting from the ups and downs of vintage quality less acutely than the winegrowers and frequently provided them with financial support. Occasionally, winegrower debt resulted in partial or total sale of the estate to the *négociant*, who became significant estate operators. Thus although innovation and détente evolved within the value chain, to a great extent the advantage remained with the *négociants*. The chateau and estate bottling were the mechanisms the winegrowers used to redress this imbalance.

In the first half of the 19th century the concept of chateau began to supplant the term *cru* and gave Bordeaux's designation of *terroir* an estate-based distinction. A chateau not only communicated "nobility" to consumers, but also the link between *terroir* and wine in a manner not done for other agricultural products.[10] The chateau evolved in Médoc, where vineyard scale and wine prices supported opulent edifices, but as the term was democratized, simpler buildings and estates gained noble title. From a handful of chateaux before 1850, 1600 had come into existence by 1908.[11] Chateau bottling began to diffuse, allowing winegrowers to combine identity and a guarantee of authenticity and many began direct sales in Northern France, Belgium, Germany and the Netherlands.[12] In 1904, the Syndicat des Grands Crus Classés du Médoc pushed members to increase bottling and to adopt standards, but this was resisted by the *négociants* and with internal weakness failed.[13]

Sophisticated use of estate-bottling arrived in the 1920s with Philippe de Rothschild. He insisted on bottling and marketing all his wine, and commissioned artists to create his wine labels. Although Rothschild organized the Médoc first growths to do the same,[14] few others could follow because, although willing, they lacked the scale of the Médoc properties. Rather, in the difficult 1920s and 1930s, the evolution of the estates was limited and cooperatives were the major response to distribution chain control. Originally developed in the south, Bordeaux adopted them rapidly during the depression, when merchants were disinclined to buy at a reasonable price or at all.

Alongside the value chain, the struggle for control of market power took place within the regulatory theatre. As an early example the Bordeaux *parlement* (many members were proprietors) prohibited importing foreign wine in 1758, despite merchant opposition. In 1764 another law stipulated stamping barrels with the producer's seal.[15] More laws protecting the origin of wines, for proprietors and territories, were introduced and enforced in the first half of the 19th century within France.[16] In 1884 the government authorized and initiated the formation of winegrower *syndicats*. These actions gained international recognition with the Madrid Agreement of 1891. Most importantly, a crisis of oversupply and escalating production costs at the turn of the 20th century increased the perception of merchant blending practices as fraud and precipitated the creation of the world's first appellation system.

Bordeaux's boundaries were defined early in the appellation system's gestation in the first decade of the 20th century, notably excluding the wines of the high country. Regional organizations of winegrowers achieved these boundaries,[17] while the merchants, through the Chamber of Commerce (supported by the national and Midi winegrower associations), defended blending, claiming Bordeaux to be a type of wine. Even the winegrowers, aware that their wines needed touching up for some markets agreed not to push for formalized regulations against blending, if the merchants practiced reasonable discretion.[18] New laws brought in at the end of WWI allowed formal horizontal differentiation by territories, but rampant proliferation of appellations threatened until the reformulation of the AOC system in 1935 (more on the origins of the AOC in Chapter 4).

Parallel to creating the AOC, winegrowers and merchants recognized the need to mediate their interdependence, particularly in regard to bulk transactions. In 1922 the merchant and proprietor Fernand Ginestet organized the *Union des Propriétés et du Commerce* to protest a luxury tax on wine. Over 30,000 winegrowers and merchants demonstrated in the streets. Later in 1930, in recognition of the need for regional government influence, the *Comité Départmental des Vins de Bordeaux* (CDVB) was set up by the *Conseil Général de la Gironde* in 1930. However, significant mutual action did not take place until the end of the second war, when the Conseil Interprofessionnel du Vin de Bordeaux (CIVB) was reborn from the CDVB. Even then, antagonism between the merchants and winegrowers, a mutual fear for loss freedom, and the need to impose fees, prevented self-financing of the institution, the power to impose quality controls, and professional certification.[19] The state stepped in, creating an institution with limited power, but enough funding to engage the two sides and for promotion of the region.

The struggle between the two sides would lead to recurrent crises and reformations of the CIVB. The first occurred in the early 1960s and led to the Bordeaux Wine Protocol. This declaration established a balance of power in the CIVB and empowered it to impose quality control and register transactions on bulk wine. Another crisis followed in the 1970s and helped to push more winegrowers into an already expanding estate sector.

In post war conditions estate wineries flourished. Mechanization was introduced for plowing, pruning, and harvesting, while pesticides and other chemical treatments became more dependable. These advances gave winegrowers time to learn about improving quality and to take more control over the elaboration and marketing of their wine. Institutional support began to give winegrowers the scale advantages held by the merchants. The *Crédit Agricole* financed equipment and short-term operating costs[20] and Bordeaux's faculty of oenology and other training institutes spread knowledge and educated the children of many chateaux.[21] Increasing urbanization and discretionary affluence in French and export markets supported the growth of the estates. Consumers looked for purchases offering distinction and pleasure, not only a resurgence in the UK high-end market, but also among Northern Europeans, and foremost among wealthy Americans reviving from prohibition. The media played a large role, particularly critics. The upper-class tone and focus on the Grand Cru of British writers gave way to Robert Parker's grading and other systems. All writers increased accessibility and introduced a far greater number of chateaux, while tellingly organizing their work according to territories and introducing consumers to them. Of note, however, is that even Parker's 1,500 estate compendium is a fraction of Bordeaux's variety.

**Management of Diversity**

Charles de Gaulle once queried, "How can you govern a country in which there are 246 kinds of cheese?" The Bordelaise ask how can you run an enterprise that has 10,000 CEOs? Their estimation is based on the 10,000 winegrowers, but it should also include the merchants, brokers, all the suppliers to the industry, hotels, restaurants and all other businesses reliant on wine. It could also include the governments, all the industry associations, schools and other institutions that want a say in how the territory is run. If Bordeaux were a company it would have sales of over €3 billion (US$4.05) and directly employ 35,000 people in the winegrowing-brokerage-merchant value chain, plus another 21,000 in related tourism, machinery, supply and other industries. In terms of total sales Bordeaux remains ahead of Constellation Brands, the world's largest corporate wine seller, at €2,846 billion (US$3,843 billion; 2007). In contrast to Constellation's few hundred brands produced in different points around the globe and servicing relatively proximate markets, Bordeaux offers 12,000 brands originating in one location requiring distribution around the globe. This is a unique organizational challenge.

The challenge is compounded by the diversity of chateaux and merchant types and the conflicts of interest among them – differences among the Grand Cru, wide variations among the thousands of estates bottling their own labels (*petits chateaux*), the bulk producers, and the cooperatives. The merchants range from large multinationals with their own brands and estates to individually run operations representing a short list of chateaux. Within any typology, there are successful and struggling firms, each with differing reasons for success or lack thereof. Differences among territories, those well known and others less so, increase complexity. Chateau bottling dominates in the former, and bulk production in the latter. Conflicts arise within territories between generic producers and estates that are trying to elevate their own and territorial reputations.

Despite their many differences Bordeaux's firms are linked structurally in several ways. The *petits chateaux* and Bordeaux brands enjoy the aura of the Grand Cru, while the Grand Cru stand out as the best of Bordeaux. Merchants and bulk producers, and estate bottlers selling bulk, are interdependent. Merchants own chateaux, while winegrowers are becoming merchants of their wine and representing other chateaux. Intermarriage famously linked the families of Bordeaux's wine masters among each other and the chateaux[22] and the same mechanism ties together merchants, chateaux, cooperative members, and brokers across all levels and territories. Indeed, all these professions are often represented in one family. Yet in addition to the benefits of these horizontal and vertical linkages, greater reputation effects and economies of scale are available if their territory is managed well.

The first place to look for management is the CIVB, whose formal mission is to bring winegrowers and merchants together for the governance of the value chain. It does not directly intervene, but develops mechanisms to allow the two sides to transact more efficiently. It also provides economies of scope and scale in R&D, marketing, and quality control and is the chief lobbyist and coordinator of the territorial system. The CIVB uses many corporate management approaches used in corporations, but is limited in its powers by the independence of the merchant and estate businesses, and by the autonomy and self-interest of each territory. It also has to contend with democratic governance of 35 representatives from 10,000 wine properties (approx. 8,000 proprietors and 22,000 employees) and 35 representatives from 400 merchants (approx. 7,600 employees). These representatives are put forward by the federations of the winegrower and merchant associations. Another 32 representatives from governmental, non-governmental organizations, and companies participate, but don't have voting rights. Among these external voices the brokers, with an association of 130 members, has the greatest say as its president is part of the CIVB executive.

Members meet four times a year to set policies, pass the budget, elect a president and other decisions. The two sides exchange the presidency every two years and also share leadership of the technical, economic and marketing commissions. A president will usually spend a few years leading a commission before being chosen, and commission leaders have usually demonstrated their capabilities in their respective associations. The Bordeaux and Bordeaux Superieur *syndicat's* representation of close to half of all winegrowers results in frequent selection of the president from its ranks. The leaders in the CIVB and *syndicats* serve as Bordeaux's representatives on the regional INAO commission and the CNAOC. The opinions of the same leaders are found in government investigations of the industry,[23] trade journals or voiced at trade events.

The commissions do most of the territory building. The economic commission monitors and publishes production, prices, sales in different markets, and advises on the imposition of sales controls. The technical commission instigates research on viticulture and vinification, diffuses technical knowledge, undertakes quality control, and investigates regulatory, environmental and territorial issues. Marketing is the most expensive and divisive of the three activities. These activities will be described more subsequently, but it can be noted that only about 30 administrative staff, operating on a budget of €28 million do the actual footwork. The modest staff and the budget, compared to Bordeaux's €3 billion production, is an obvious difference with the vastly larger bureaucracies of similar scale corporations. The funding is provided by a government authorized extraction of fees from both winegrowers and merchants.

The CIVB is not, however, the sole source of guidance for the territory. The FGVB (*Fédération des syndicats des grands vins de Bordeaux à Appellation Contrôlée*) federates Bordeaux's 35 winegrower associations and provides a stronger regional voice for direct and indirect influence on the CIVB, the national federation of winegrower associations, and the national committee of the AOC (CNAOC), and unto higher political bodies. It facilitates information sharing, but doesn't control the territories. The two merchants' associations are linked into a federation, and provide educational and political functions, but without territorial differences to smooth over. The INAO, from its regional office supports and monitors the CIVB and the FGVB. The departments of External Commerce, Customs, Taxation and Fraud Prevention help the CIVB manage the territory by reducing the cost of monitoring transactions and market research. They also regulate and tax winegrowers and merchants. Several other departments, not directly linked to the AOC govern the industry. The Ministry of Health, for example, has installed itself as an overseer of the system and as a *bête noir* to most winegrowers.

The Region of Aquitaine and the Department of the Gironde provide a range of programmes for the industry and a diversity of specialist organizations in research, education and training, marketing, organic agriculture, environmental protection, and tourism and regional promotion have evolved. These organizations contribute policy direction, integration and economies of scale and the CIVB tries to coordinate their efforts. Ultimate sanctioning and subsidy of the system depends on the French and EU governments. They fund education, research, infrastructure, and other direct and indirect support, particularly when entreated by regional stakeholders. The CIVB lobbies for French and EU support of appellation protection at the TRIPs and other international trade discussions. The EU, of course, supports the wine industry, through the common agricultural policy and provides subsidies for excess wine distillation and vine-pulling. Bordeaux has long availed itself of the former and recently of the later. Of course other policies impact the industry and require a response from Bordeaux.

*From Bilateral Mediation to Pluralistic Governance of the Distribution Chain*

Bordeaux's variety depends on a corresponding variety of distribution chains. In bilateral relations a winegrower sends most wine or grape production to a merchant blender or cooperative. In so doing the winegrower avoids distribution and marketing costs, but loses control over differentiation and the potential for higher returns. Bilateral relations still generate variety through numerous territorial, stylistic and merchant/cooperative brands. Multi-channel relations amplify variety by helping estates reach a sufficient number of customers and retain differentiation, but requires more work and imposes greater costs. Territorial management opens up the number and diversity for both types of distribution channels, but several issues require management. Information asymmetry has to be reduced to provide transparent bargaining between winegrower and merchant and preclude domination by one or a few merchants. Stability and fairness in terms of trade is necessary to enable both sides to invest in improvements. Production and price stability avoids flooding the market, destructive competition that drives prices down or up too high and the inciting of opportunism among winegrowers and merchants. Neither side, nor segment of distribution chain relations, can be allowed to damage Bordeaux's reputation. Nor can the distribution chain impose excessive double or triple margins on consumers, thereby allowing corporations the opportunity to internalize transactions and eliminate channels.

The present structure of the CIVB resulted from the boom and bust of the early 1970s. Markets had expanded in the 1960s, especially the US, generating rampant speculation on a limited supply. Prices soared in all categories and both winegrowers and merchants capitalized. But when prices soared too high, and the 1973 vintage failed, buyers looked elsewhere. When markets crashed each accused the other of greediness for unsustainably escalating prices. Merchants complained that each grower demanded more upon hearing rumours of how much their neighbours were asking.[24] Winegrowers took to the streets and withdrew from the CIVB. They believed the merchants and the CIVB not only failed to act, but hindered their efforts to mitigate price destruction.[25] Bordeaux's reputation was further tarnished when one of the oldest merchant houses was caught using non-appellation wine in its blends.

Similar events occur repetitively. The excellence of the 1995 and 1996 vintages prompted a run-up in prices, which initially inflated the poor 1997 vintage. The subsequent deflation damaged the reputation of winegrowers, merchant's and Bordeaux. In a drop in demand after the turn of the millennium, bulk wine winegrowers blockaded the CIVB's doors demanding action on prices that had fallen below the cost of production.

The crises are in fact structural, derived from a lack of knowledge of production volumes and market demand, and of transparency in price setting and contractual uncertainties. The winegrowers encounter the situation acutely as an information asymmetry caused by distribution and market stabilization residing in the hands of the merchants, with the plurality of a merchant's buying and selling giving them an advantage. Quality differences among the vintages provide ample openings for speculative opportunism from both sides. In 1976, the CIVB was reformed and given the authority to provide direct and indirect governance over the bilateral relations of generic production and distribution. These steps were necessary to bring the winegrowers back into the fold, and as these transactions still represents half of Bordeaux's volume, it remains a central focus.

The ability to negotiate the binding three-year contracts was the highpoint of direct bilateral governance. It provided stability and planning capacity for winegrower and merchant. EU anti-monopoly regulations eliminated this tool, however, and returned winegrowers to information asymmetry between buyers and producers and a short-term commodity-pricing regime. This loss of power has been described as a critical weakening,[26] but other mechanisms, while less direct, perhaps more powerfully support the territory's diversity. The CIVB was empowered to register all transactions and publish a summation of average prices for barrels of bulk wine, thereby reducing information asymmetry. The transaction statistics support territorialization because they show differences in prices by appellation and indicate whether strategies used in each appellation are paying off. The CIVB can also still compel producers and merchants to withhold sales, the so-called "qualitative reserve," when it reckons too much wine is being marketed from an appellation. The rule of thumb is that merchants and winegrowers should have a ratio of stock sold through the year to stock available at the end of the year of approximately 1 to 1.5.

The CIVB's attempt to stabilize supply and prices is a supplement and balance to a key territorial role of the merchants. They can blend and bottle 4.5 million hl of wine (more than half Bordeaux's annual production) and store 3 million hl. That capacity, however, is also a primary advantage over the winegrowers. Maintaining stock imposes costs on the winegrowers, but it also puts them in a stronger position, and is a discipline that drives them to bottling. As an innovative middle road, chateaux and merchant collaboration on bottling for chateau labels gives a merchant a guaranteed source whose quality they can foster and the winegrower a degree of differentiation, but without the costs of distribution and marketing. Despite the amelioration of bilateral relations, particularly since the crisis of the early 1970s, winegrowers have recognized the most effective way to build the value of their wine, even if they have to put more effort into it, is to bottle their production and find several distributors to sell to a global market.

The medium for the shift to multi-channel distribution, and crucial for the flourishing of the chateaux, was *la place de Bordeaux*. It is the territory wide market place where, with the brokers' mediation, the variety produced by the winegrowers is exchanged into the merchants' hands. All the merchants, brokers and winegrowers, whether transacting in bottles or bulk make up *la place*, but the system works best for one of the region's more remarkable institutional innovations. *En primeur* sales are a futures market, not for fixing prices on commodities, but for individual chateau to foment competition for their differentiated product. The system originates in the centuries old practice of merchants buying wine the spring after harvest (*en primeur*) and then transporting it to their own cellars for aging and blending. When chateau bottling proliferated from the 1960s the system transformed itself whereby a chateau would offer tastings to merchants, importers, and wine critics in the spring. This helped set their prices and sales a year before bottling, assisting cash flow tremendously. Where chateaux used to sell to a limited number of powerful merchants, they could now sell to a greater number of merchants with diverse global customers. Furthermore they could reserve or withdraw allotments year to year to encourage loyalty and effort in selling to consumers and stagger sales through the year to capture any upward speculation in price.

The transformation of *la place* certainly doesn't preclude opportunism, just balances it. The power shift occasionally prompts merchants to complain of price gouging, disloyalty, and inability to serve customers properly. Producers may force merchants to buy poor wines that they have to sell at a loss or buy their second, third wine, or even wines from another chateau. The chateaux counter that they need a diversity of channels for a volume that despite being small requires a wide distribution to obtain customers. Furthermore, the merchants may fail to distribute at the prices and to the customers the winegrower wants in order to build his sales and reputation. Instead the merchant may hold on to wines or sell them elsewhere for greater gains. Despite these problems, the system enhances the territorial linkages because, to ensure access to better-known chateaux, Bordeaux's merchants buy good and bad vintages and find some means to sell both. In comparison external buyers pick and choose their years. The merchants of *la place* distribute 75 percent of the regions wines to over 160 countries.

The evolution of *la place*, particularly *en primeur* sales, was driven by the Union des Grand Crus (UGC), an association of 120 top chateaux, predominately from the left bank. The association organizes tastings (for 1,500 in 2000); invites journalists, critics, and merchants; puts them up in hotels; and entertains them lavishly at famous chateaux.[27] The event is played up by the trade and media, thus not only does the UGC use the territory for marketing purposes, but it turned the market into a marketing tool. The UGC, however, represents a small fraction of Bordeaux's total volume, and others want to join the act. The *Cercle de Rive Droit* provides renowned chateaux on the right bank (and those seeking renown for wine they feel is undervalued) *en primeur* services. Many territories operate *en primeur* events and some chateaux independently offer tastings. Throughout Bordeaux, *syndicats* and organizations such as *Vigneron Independents* set up venues to capture the wine trade as it descends. About 5,000 people attend the *en primeur* festival, expanding the event's influence well beyond the most renowned wines.[28]

Thus although only a few hundred top chateaux gain the greatest benefits from *en primeur* sales, the *petits chateaux* that represent the majority of estate-bottlers gain from the event, and more importantly from the establishment of a multi-channel standard. For example, the *petits chateaux* in the more renowned appellations use *la place* almost exclusively. Chateaux in less reputed appellations have less leverage with the merchants. They focus on direct sales to private customers, cafés, hotels, restaurants and specialist wine retailers, but eschew the supermarkets and their demand for large volumes and lower prices. To succeed in these markets, many turn to the *Vigneron Independants*, an association of independent chateau producers that provides marketing advice, set ups salons to meet customers, and some economies of scale in materials.

The CIVB has supported the rise of the estates, but at the same time is challenged by the greater complexity. Its bilateral tools are of limited use. The registration of transactions only renders information on bottle shipments and no price information (except for bottles for merchants). Average market prices are of less relevance to pointedly differentiated products at any rate. Even market timing varies as the price of most Bordeaux peaks in the spring wine trade fairs, while *en primeur* prices often rise with sales of latter *tranches* at the end of the year and end of summer. To open new markets the CIVB works with the supermarkets on wine fairs that increase exposure for hundreds of chateaux that would otherwise see little shelf time. Bordeaux's size and ability to coordinate its diversity gives it leverage in a supermarket monosopony that generally dismisses producer variety in order to reduce transaction costs.

Bordeaux's size and coordination also enabled Vinexpo to become the world's largest gathering of wine sellers and buyers. Established in 1981 by the Bordeaux Chamber of Commerce and Industry, along with help from SOPEXA the French export promotion agency, it draws the world wine industry to Bordeaux, and on an alternating basis it takes Bordeaux (and others) to the Americas and Asia. There are other wine fairs, notably VinItaly and the New York Wine Experience, but despite being national fairs they are not the size or importance of Vinexpo and are not operated by a territory.

*Reputation and Marketing: From Tradition to Modernity*

Any winegrower or merchant will tell you that the name Bordeaux opens doors around the world. Their opinions are born out by their own surveys that found 80 percent awareness of Bordeaux in northwestern Europe[29] and independent studies showing 94 and 80 percent awareness in the UK and US.[30] Other studies have used Bordeaux as proof that building reputations over the long term benefit both territory and chateau.[31] The difficulty of building, the maintenance, and the importance of these reputations cannot be taken for granted, especially in the face of global competition.[32] From the outside, the foundations of Bordeaux's reputation, its chateaux, grapes, blending styles, appellation structure, classification systems, even its bottles, are deemed traditions. But these are consciously designed mechanisms of differentiation, which are increasingly being buttressed with modern marketing tools as the Bordelaise adjust to compete with the approach of new world producers.

Variety shapes all reputation building and marketing efforts. Collective action is required because of the essential tension between the umbrella Bordeaux label and the differentiation of 57 appellations and 35 territories and the vast number of chateaux. The varieties, blends, techniques, *terroir* – the *typicité* of the region requires a coherent package to break down information asymmetry with consumers.[33] Bordeaux's success is demonstrated by its position as the global benchmark for the Bordeaux blend of cabernet sauvignon, merlot and cabernet franc, and indeed even for cabernet or merlot varietals. Among the appellations, territories and chateaux there are, however, characteristic blending styles and an offering of renowned and less renowned whites, sweet wines, rosé and clairet. Variety is amplified by the chateaux and merchants who bottle first, second and third wines and as unique chateaux or brands for different markets.

A cliché, oft repeated by merchants and winegrowers is to claim they don't know all the appellations, so how can they expect this feat of understanding from Americans or even the French? Yet if recognizing the appellations is difficult, making sense of 12,000 labels without some guides would be next to impossible.

The territorial recognition once derived from the Bordeaux (or Burgundy) bottles has been weakened as serious winemakers choose these shapes to signal their ambitions and mass producers capture the premium it denotes. The tag "Bordeaux blend" implies a similar de-territorialization. The US "meritage" society has even re-branded the blend, but while enabling producers to avoid the limitations of the varietal system of denoting quality, the label can be used by any winegrower, anywhere. More resilient to such appropriation and critical to maintaining long-term reputation and balance between coherent signal and diversity, Bordeaux has structured its appellations and classifications. The latter, while market-based configurations unique to Bordeaux, are discussed in the next chapter as they are controlled by the territories.

Appellations hierarchies are used for both red and white wines and to a certain degree overlap, except a division of the top red and white appellations. The hierarchies are used to guide consumers to wines of different prices and qualities, much in the way a company offers a range of wines. The Médoc communes and Saint-Émilion sit on top of the red wines, followed by the broader Médoc and Graves appellations, the five côtes take the lower middle, and generic Bordeaux is stuck at the bottom. Sauternes represents the pinnacle of sweet white wines, but the best dry whites are less distinct, although Pessac-Leognan and the Graves might claim that distinction. This hierarchy is semi-officially based on *terroir* differences, but is also technically written into differences into yield limitations, vine spacing and other regulations that govern appellations. The structure has a historical reference in assessments by the brokers dating to the 17th century, but the INAO had a strong hand in building this structure as it guided the definition of regulations in the appellations.

The hierarchy's distinctions are subtly portrayed in the publications of the CIVB and indeed are ingrained in Bordeaux's culture. Unfortunately for many of the lower appellations, particularly the côtes, this hierarchy doesn't recognize internal differences in *terroir* and evolution in winegrower skill. The appellation regulations were written when winegrowers in many territories were not as organized or financially and technically capable. The market and wine critics, however, recognize improvements and many winegrowers, who formerly only produced bulk wines, now chafe at the hierarchy, and what they believe is a merchant controlled CIVB maintaining it. *Syndicats* also work to improve their standing.

Extensive marketing campaigns developed from the 1980s with the realization that traditional mechanisms of differentiation no longer sufficed. From the outset, and increasing subsequently, the CIVB has balanced generic brand building and promoting differentiation in its marketing activities. Most winegrowers and merchants have always used Bordeaux on their labels, but in recognition of the power of brands and logos, the CIVB voted to make this a requirement. They have also tried to update the typically used "Grand vin de Bordeaux" with a stylized, but optional, "B" logo. To reduce proliferation of labels, each chateau is limited to the use of two chateau names. The umbrella brand strategy is pursued through mass advertising. For example, the advertising campaign "seduction" used billboards, TV and radio clips, magazine ads, tabletop guides in restaurants, internet ads, TV and radio, pizza boxes, etc. to convey its image in a few dozen foreign and domestic markets. On the other hand, tasting events, held at supermarkets, expositions and other venues, are used to entice consumers to move into higher valued wines. They focus on differentiation by appellation and chateau, particularly mid-range chateaux. Twenty percent of the CIVB's marketing funds are used for the six appellation families (*groupes organiques*) and spent on trade events or supermarket promotions where the territories and other groups such as the Union des Grands Crus are invited to join.

The marketing strategy also reflects a balance of interests and attention is paid to which chateaux are chosen as representatives. Large volume brands and grand cru wines don't receive much attention, under the assumption they can take care of themselves and benefit from association with the image advertising. A commission tests the wines both to provide fairness in selection of chateaux that will get this opportunity and to preserve the regional reputation.

The CIVB devoted 80 percent of its budget to marketing in the 1990s, then in recognition of heightened competition, voted to raise dues to double marketing in the early years of the millennium. The CIVB's marketing remains less than 1 percent of bottle costs, on the face of it a low figure in comparison to corporations that spend 10–15 percent. The dues are also graded so that higher valued wines pay more. The increases were not universally popular, especially among the generic producers and the petite chateaux. Despite grading, prices paid for a grand cru mean that their dues will be a smaller proportion of total costs than that paid by a lesser-known chateau. More importantly, the individual producer cannot link the CIVB's expenditures directly to their sales increases, while they still have their own marketing and distribution costs. Added to that was skepticism of the marketing campaigns.

On the other hand, the CIVB tries to overcome the inability of producers and distributors to analyse different geographical and demographic markets, the sales outlets, what appellations and styles of Bordeaux are being sold and for what reasons. It analyses information gathered from the registrations of sales and from the customs and tax departments. Working with the authorities also enables the CIVB to collate reporting and reduce some red tape for the merchants and producers, and incidentally keep track of the fees owed it. Surveys of consumer trends are also purchased, as well as information on international markets. The research is disseminated through a variety of venues on a yearly and monthly basis.

The collective marketing of variety exacerbates winegrower and merchant tensions because the opportunism inherent to multilateral relations means that neither can be guaranteed direct returns when sharing marketing costs. Whether a system to share marketing costs can be devised is an open question, but it is rarely practiced in Bordeaux. Merchants regard themselves more as distributors than marketers, and few engage in it. For the most part the chateaux must build their own reputations and use territorial reputations and classifications to undergird their sales. That provides them leverage with the merchants or in direct sales. Merchants use those reputations to sell their portfolio and usually reserve their marketing efforts for their own brands.

*Production and Quality Control and R&D*

Territories, appellations, classifications, and chateau labels enable a great deal of differentiation within the territory, but these efforts can be undermined if Bordeaux's overall reputation is not sustained. In the 10 years bordering the millennium, the quality control of generic Bordeaux and many bottling chateaux compared poorly with many new world brands. As Bordeaux's reputation fell, so did the sales and prices of all but the Grand Cru. The divergence generated soul-searching in the territory, and Christian Delpeuch, the then (merchant) President of the CIVB, famously declared over 30 percent of Bordeaux to be undrinkable. The problems were deep seated, including the use of inferior *terroir*, over production, and freeriders shirking quality control. Yet they were not unrecognized, rather the quality and production control of 10,000 producers stimulated a complex set of measures, including widespread investments in training and R&D to improve the level of quality and reinforce differentiation.

*Production control* Bordeaux has had an ambivalent approach to imposing the limits on production that create a monopoly over the appellation and rarity value. Vineyards in the Gironde have increased 6,000 ha since 1960. More importantly, until the mid-1970s, about 30 percent was made into white table wine and not Bordeaux AOC. Through the 1980s almost all this production was converted to red AOC Bordeaux. Production rose from 3.7 million hectolitres in 1962 to a capacity of 6.5 million hectolitres in the early 2000s, again with much of the increase occurring in the 1980s. This increase occurred in the areas producing Bordeaux generic (10,000 ha between 1993 and 2002 for example), but there were also notable increases in the Médoc (2,000 ha) and the côtes (4,000). This increase was absorbed during the expansion of the red wine markets of the 1980–90s, but by the late 1990s the quality of generic Bordeaux was of concern and consequently demand and prices dropped significantly.

Expansion and conversion of vineyards was possible because although the *syndicats* and INAO are supposed to restrict AOC land, for many territories a considerable area was deemed within standards. Delimitation took place at the outset of the AOC system, and reconsideration of vineyard delimitations has been a difficult process with limited results. The geologically inspired INAO can push for changes and must arbitrate any requests for expansion, but only the *syndicats* can initiate any re-evaluation of their vineyards. Bordeaux as a collective can only exert influence through the CIVB, the FGVB or the CNAOC. Their efforts face the reality that a collective decision must be obtained in the *syndicats*, a decision that will deprive some owners of the use an appellation and cause a loss of income. Such a change also imposes the costs of analysis before the decision and further transactions costs after: the latter including legal appeals, changing cartographic records in commune offices and so on.

Alongside AOC land control, the EU limits the planting of new vines in order to reduce its wine lake. Each country is allotted a quota, which they then dispense internally for allocation to winegrowers. The FGVB runs an exchange for these rights in Bordeaux, and for planting in Bordeaux and the pulling of vines elsewhere. The system frustrates winegrowers who want bring good unexploited land into production without unnecessary costs.

*Quality control*   The limited ability to control expansion and conversion of vineyards to AOC was a mixed blessing. The flexibility allowed many quality chateaux to come to the fore, in spite of appellation and classification hierarchies. On the other hand it allowed into Bordeaux and other appellations, thousands of hectares of vines, whose spacing and other viticultural practices, belonged to an era when vignerons tried to "make the vines piss." Many problems receded with the national trend to the consolidation of agricultural land. The still remarkable number of 10,000 winegrowers is down from 47,000 in 1960 and the drop-off is expected to continue, but the need to deal with quality disparities remains.

The *syndicats* are the foundation of the QC system used throughout the AOC and Bordeaux's *syndicats* developed these systems. Proximity and competition encouraged rapid adoption by the better-organized *syndicats* in the 1950s. They set an example for merchants and winegrowers concerned about the reputation of generic Bordeaux to establish a QC system in the 1960s through the authority of the CIVB. Authority for that quality control now rests with the Bordeaux and Bordeaux Superieur *syndicat*, but the CIVB still supports quality improvements. For example, with the INAO, it has worked with the *syndicats* to lower yields and as winegrowers require flexibility between good and bad vintages uses a system that allows yield fluctuations as long as a 10-year average is met. According to some *syndicat* leaders, the experiment has been somewhat confounded by the INAO's continued micro-management of yearly reports.

Unable to enter the *syndicat*'s upstream quality control, the CIVB samples bottles in distribution, testing and grading them, then gives the results to merchants or winegrower *syndicats* for use. The system sorts out whether problems arise in production or distribution and clarifies accountability for both sides. However, it is up to the *syndicats*, to decide whether they will use the system. Some such as the Médoc make it an important part of their QC systems, but others remain unconvinced that responsibility in the system can be discerned. One *syndicat* does not use the system with confidence because its winegrowers think some merchants exchange their wine for that of more prestigious appellations, and bring recriminations for poor performance onto them.

New world competition compelled the CIVB and ONIVINS, to develop the Plan Bordeaux to get production and quality under control. Unprecedented, for Bordeaux, its vine pull scheme targeted the removal of 8–10,000 ha, or close to one-tenth of the vineyards. Additionally, a *vin-de-pays* category was to provide an outlet and remuneration for wines that should not go into Bordeaux, and the limitation of yields per hectare. Delpeuch was the force behind this plan, but resigned his Presidency of the CIVB in 2006 when the government wavered on compensation to winegrowers who pulled their vines and to opposition from some quarters of the winegrowers. In the first years, the vine pull scheme barely achieved 20 percent of its target, but the vin-de-pays d'Aquitaine was created and the CIVB voted to limit all yields in Bordeaux to 50 hl/ha.

All French wine production was feeling similar competition and demands for reform. Systemic changes to the AOC system would of course influence Bordeaux. The INAO's first attempt, under René Renou, was to broach a system of *AOC d'Excellence*. The system recognized elite *terroir*s, but foisted more information on the label. Bordeaux rejected this idea because of its already complicated appellation and classification systems and the degradation of *syndicat* autonomy. The INAO responded with reform of the *syndicats*, changing their name (*organismes de défense et de gestion*) and activities. The key change is replacing the voluntary taste-testing panels with a third party organization, thus addressing a critique of AOC quality control. Bordeaux as a whole accepted this change, brought in while the INAO streamlined the operations of its national committee and administration.

### R&D and Training

A more profound foundation for quality, and perhaps Bordeaux's greatest advantage, is the network of training and research institutions that provides broad and deep knowledge of the technical, managerial and cultural aspects of the industry. They serve the winegrowers and merchants, and the diversity of tourism, laboratory, machinery and other specialists involved in the industry. Many precursors of these education and training institutions existed before the war, but their diversification and diffusion coincided with the conversion of chateaux from bulk to bottles.

The agricultural Lycée at Blanquefort is France's largest centre for viticulture and viniculture research, offering diplomas, apprenticeships and continuing education. It is responsible for educating successors to chateaux and the increasing ranks of salaried professionals. The agricultural chamber of commerce provides continuous upgrading through its network of six oenology centres and another dozen offering agricultural advice. The Lycée and Chamber are complemented by many other public, private and non-profit organizations that provide training in everything from pruning to marketing. The CIVB also disseminates basic and innovative technical information and provides weather, pest and disease monitoring. Many *syndicats* adapt this information to their local requirements and provide consultation services to their members.

Bordeaux's status as the benchmark for fine wines does not rest on the transmission of skills and traditions, rather extensive research and innovation. From origins in the 19th century, the Faculty of Oenology at Bordeaux University 2 has immensely influenced Bordeaux's proprietors, merchants and consultants. It has led the world in improving fermentation, introduced malo-lactic fermentation, and perhaps most notably, improved the linkages between vineyard and wine cellar.[34] The faculty's activities are matched by research at the Bordeaux branches of the

National Institute for Agronomic Research (INRA), the National Agricultural Technology Engineer's College (ENITAB), the interprofession's Institute for Vine and Wine Technology (ITV), and other university departments. Like the faculty of oenology, they mix basic and applied research, covering vine physiology, analysis of fermentation and aging, pests and vine diseases, and chemical verification of origin. Other institutes investigate the historical, geographical and economic basis of the region.

Research has been marked by close association with winegrowers and merchants. Peynaud, the most influential academic, was drawn into the Faculty of Oenology from consultation. He made scientific understanding of winemaking accountable to taste, moving oenology back into the vineyards and an insistence on the health of the grapes. Today the chateaux and the faculty are interdependent, indeed the faculty yearbook resembles a wineguide. Research collaboration extends to the prevention and prosecution of fraud, linking customs, the fraud department and the INAO. The several institutes and researchers provide for rivalry within the territory, with contributions from national and regional offices of the INRA, ENITA and ITV. In the 1990s the Aquitaine Regional Council moved to bring these efforts together by establishing the Institute of the Science of Vine and Wine to create a global research centre. With support from the Department of the Gironde, EU, the Ministry of Research, the Urban Community of Bordeaux and the CIVB, by 2009 it had evolved into Europe's (and likely the world's) largest viticulture and oenology institute.

Sustaining Bordeaux's unique territorial identities is an important part of this research, and accomplished in close collaboration with winegrowers and funding from the CIVB and the *syndicats*. Research has investigated characterization of *terroirs*, appropriate rootstock, and the unique qualities and care needed for Bordeaux's grape varieties and vine diseases. The CIVB runs a colloquium every two years to publicize this type of research. Perhaps the most important collaboration is information sharing among winegrowers, as most practical innovation originates with them. Sorting tables and the green harvest are good examples of chateaux owners innovating and diffusing advances within their territories and throughout Bordeaux.

*Territorial Development and Defence*

Bordeaux has entered a more complex era requiring a sophisticated ability to coordinate through the use of soft power. It has to gather the resources and coordinate efforts to protect and build physical and cultural infrastructure, enhance the environment, provide pensions and healthcare, and deal with social and political demands for changes that threaten the industry. On several of these issues the interests of the wine territory confront those of the municipalities within it and tangle with diverse agendas at the national and EU levels. Winegrowers put more faith in their industry representatives than representatives elected for the various levels of government.

Bordeaux may be famous for wine, but most of the Gironde's 1.3 million, residents are employed in services and manufacturing industries that have eclipsed wine's leading role in the regional economy.[35] Bordeaux's main economic role has long been as a regional government, commerce and transportation centre.[36] In the 1960s and 1970s automotive, electronic and aerospace transformed the suburbs, and more recently, new biotech and ecotech parks and a laser route are extending the periphery. While the population of central Bordeaux (commune) declined from 278,403 in 1962 to 235,878, the 27 communes of Bordeaux Urban Community rose from 497,807 to 714,727, and the greater metropolitan area Bordeaux-Arcachon-Libourne from 564,896 to 1,010,000.[37] The area is also growing faster than the rest of France selling itself as both a high tech agglomeration and a capital of good living. The employees of these industries construct homes within and among the vineyards, and are never too far from one of France's formidable hypermarket complexes. Internal and external connectivity increased as Bordeaux developed into a hub of the EU highway and train systems, the latter being upgraded to LGV in 2016.

Much of the industrial location that fuelled suburban growth was assisted by the four decades of Jacques Chaban-Delmas' mayorality and influence at the national level (1947–95). He was trying to compensate for the withering of colonial trade that once made Bordeaux the most important port and also for the failure of Bordeaux's merchants, including those of the wine trade to expand their investments beyond their specializations.[38] The consequence of development, however, was the extinguishing of the vineyards that once extended up to the city's gate. Haut Brion – the seminal chateau and brand – is isolated within a subdivision, where once it was surrounded by rival chateaux. In the near perimeter of Bordeaux city, 750 winegrowers, have disappeared for a total loss of 5,000 ha[39] – a figure that rivals any subsidized vine pulling. In the exurbs, often there is little participation from winegrowers in commune decision-making and division between municipal objectives and wine territory objectives can be pronounced. In the Médoc, politicians saw the vineyards providing little tax income, and undoubtedly galling for winegrowers, enabled the prized limestone gravel of vineyards to be turned into gravel pits.[40] In 1986 the winegrowers successfully resisted the location of a technopolis in La Brède (18 km south of Bordeaux) and more recently when a highway bypass threatened vineyards in Médoc winegrowers used the Internet to obtain 5,000 signatures for a petition of opposition.[41]

The CIVB and the FGVB have long considered much of this development as an attack on its production base and damaging to the cohesiveness of the industry and landscape imagery. The *syndicats*, and the Chamber of Agriculture, working with the CIVB, INAO and Ministry of Agriculture gained the legal right to review threats to AOC lands, including road construction, development, and urbanization plans.[42] Yet, they can only suggest changes and final decisions depend on the local politicians' evaluation of vineyard contribution to their community.[43] Because of these limitations, the CIVB works with government, the public, and stakeholders to emphasize the industry's role in regional employment (60,000) and quality of life. It has succeeded in getting protection for vineyards in urban planning, particularly through less ambiguous zoning of agricultural land, and a recognition by the planners that winegrowing remains the single most important economic force in the region.[44]

In 2001 their efforts gained recognition for viticultural heritage when a master plan for the 91 communes of the Bordeaux Metropolitan Area was designed as a systemic integration of urbanization and viticulture.[45] It was the first initiative of its type in France and carried out in consultation with all representatives of the industry. The master plan assisted the winegrowers to protect their vineyards against a powerful threat in a subsequent departmental plan quarries.[46] Local urban plans, integrating sustainable development, were initiated as well and created space for more winegrowing input, which many have taken advantage of. As mayor of his commune, Roland Feredj, director of the CIVB, tried to convince other mayors that wine estates were not solely interested in their private benefit, but provided the communes and regions with a powerful economic and cultural force.[47] As we will see in the case of Saint-Émilion, however, in some territories municipal and winegrower interests are in harmony.

Since Alain Juppe became mayor in 1996, the city and wine industry have worked more positively together. Along with the chambers of commerce and regional governments, they collaborate on Vinexpo and Vinitech (the world's leading wine trade and viticulture and winemaking technology fairs) and the yearly festival of wine (Fête le Vin) held on the quays of the Garronne. The same group gained UNESCO's recognition of Bordeaux as world heritage site and worked to establish the Great Wine Capitals Global Network. The latter promotes business development and education, along with its emphasis on wine tourism. The CIVB's headquarters contributes an important edifice to the 18th century cityscape that won Bordeaux UNESCO recognition. Indeed, after turning its back on its waterfront, the city embarked on a rejuvenation that incorporated its wine-trading heritage into the transformation of the abandoned docklands into a tourism and commerce centre. The city's tourist office is the frontline of the tourist effort, not only within the city, but also organizing tours into the chateaux and through the vineyards. The CIVB assists these efforts by providing information and contacts, and also runs a wine tasting school and a wine bar in its headquarters. However, the CIVB is limited in capacity to motivate the majority of chateaux or their *syndicats* to participate in tourist activities.

At the national and EU levels, the territory relies on the CIVB and other industry lobbyist rather than elected representatives from the region. The outstanding exception is the contribution of Gironde senators, a large proportion of whose constituency owns or are employed in the vineyards and trading houses. Gérard César recently has prompted investigations into competitiveness of the industry, and follows in the tradition of Capus, the Bordeaux Senator of the early 20th century, who fathered the AOC system. Ministries such as agriculture, customs, finance, and education, as seen in the various assistance they provide the CIVB, the INAO, and research in Bordeaux are supportive to territorial goals. But in a land where wine was once considered integral to the culture, other government forces seek to unwind the relationship.

After a period of enjoying the halo of the "French paradox" in the 1980s the Ministry of Health has pushed legislation that has pursued the French wine industry for its deleterious impact on people's health and the horrendous mortality on France's highways. By 1991 France introduced a restrictive advertising law controlling what media can be used, what times, the contents, prohibits sponsorship and requires a health warning.[48] The Evin law mandated that anything not precisely permitted is forbidden, while ambiguously allowing wine to be advertised in cafés, bars, etc., but only describing the producer without suggesting enjoyment. Post 2000, government advertising graphically dissuades citizens from drinking and enforcement became much stricter. The latter pushed on by the ANPAA (the French association for the prevention of alcoholism and addiction) which began to target the advertising of territorial organizations like the CIVB. In the difficult times of 2003–4, winegrower reaction to the Efin law intensified as the government began to combat advertising on the Internet.

Merchants and winegrowers argued for recognition of differences in consumption patterns and wine's health benefits, particularly that wine shouldn't be equated with beer and liquor. Territories also challenged the laws, but the authorities were not amused by Burgundy's overtly sexual advertising. The CIVB's more demure "drink less, but drink better" campaign faired no better despite its use of attractive winegrowers and merchants to comply with the law's stipulation for real industry representatives in advertising. It had more success giving bags to restaurants to enable customers to take home an unfinished bottle. Eventually, the various territories and the INAO used the agricultural ministry to balance the power of the health ministry and succeeded in loosening regulations for appellation wines. Advertising can use the taste, production, regional qualities and type of grape used, but cannot allude to virility or seduction. Even Sarkozy, a teetotaler who had brought in tough anti-drunk-driving laws as interior minister supported the changes, claiming wine to be part of France's culture.[49]

Intense lobbying occurs at the EU level where through the CAP (Common Agricultural Policy) and its OCM in wine (Common Organisation of the Market in Wine) directly impact on all winegrowers and merchants. Originally organized to reduce overproduction of high volume table wines, its responsibilities have been complicated with health concerns, competition from new world wines and overproduction of AOC wines. It also plays an important role harmonizing international standards on viticulture and winemaking practices through the OIV and ensuring international recognition of designation of origin and traditional practices at the WTO. Besides national bodies like ONIVINS and CNAOC, the winegrowers are represented at this level by the European Agricultural Union (COPA-COGECA), the merchants by their Comité Vins, and wine regions by the Assembly of European Wine Regions (AREV). An all-party wine group from the European parliament brings some constituent influence to the commission. It is, however, through the CIVB that most of these lobbying forces are influenced.[50]

The OCM's primary tools have been control over planting rights and subsidies to pay for the distillation of surplus wine, grub up vines and promotion. Unsurprisingly, the EU countries, particularly the Northern non-winegrowing countries do not like paying this support and have reformed the system in 1999 and 2008 and are looking toward greater reforms in 2016. Bordeaux brings mixed feelings to these reforms. Some winegrowers have long used distillation money, but only recently has Bordeaux asked their agricultural minister to ensure access to EU funding for grubbing up vines. And while many winegrowers have long chafed against the restrictions on planting rights, the territory also viewed the potential elimination of controls as an unleashing of vineyards in high volume areas and a threat to AOC style wines. Winegrowers also feared a banalization of the designation of origin concept through the loss of exclusive use of terms like vintage and the allowance of up to 15 percent non-local wine. Winegrowers, in person and through the CIVB, Syndicats, Young Viticultures, and other organizations, made their opinions clear to the head of the OCM, Martinna Fischer-Boel, when she visited Saint-Émilion in 2007.[51] With a view to further deregulation, the interprofessional bodies of the territories (Comité National des Interprofessions des Vins et Eaux-de-Vie à AOC) commissioned a report for the OCM defending the various market mediation mechanisms used by the CIVB and other similar groups.[52] The dependable Senator César also organized a commission to gather both French and Italian opinions to influence the OCM. The 2008 reforms while providing for more grubbing up, essentially left further reforms until 2016. An important exception is requiring approval of changes to approved winemaking practices, which has been left in the hands of the OIV or in other words to international governance.

Two other issues also illustrate the challenges and benefits generated by the interdependence of territorial goals with other levels of governance. In addition to preserving the landscape, there are a host environmental pressures to deal with, in the form of regulations from the EU on down. Wineries producing more than 500 hl are accountable to a schedule of impacts on air, land, water and noise.[53] Regional and local governments help deal with effluent by subsidizing facilities on or off site and suppliers help recycle packaging used in the winery. Winegrowers who bottle must join in the EU's bottle recycling systems. A number of producers, such as the influential Jean Michel Cazes, have taken it onto themselves to develop the environmental management systems that qualify for ISO 14000 and government and buyer recognition.

The CIVB, realizing the impact of both regulatory and market governance over environmental impacts is increasing emphasis on sustainable winegrowing and made it a theme of research in the 2007–10 cycle. An early outcome was a full lifecycle carbon footprint analysis for the industry.[54] The results are challenging because Bordeaux emits only 10 percent of its greenhouse gases directly from vinification or viticultural processes. Rather 45 percent comes from the production of materials such as glass and cork, the very basis with which chateaux bottle and capture the value of differentiation. Another 18 percent accrues from transportation of wines, and 12 percent from transportation of people, particularly for sales. The CIVB intends to reduce emissions by 15 percent by 2013 and 75 percent by 2050. To achieve those goals it will have to come up with a reasonable solution to the chateau-bottling problem, perhaps return to barrel shipments as in previous epochs. It will also have to achieve cooperation upstream and downstream in the multi-channel distribution system to coordinate changes in bottling and marketing. In contrast, large brands, with integrated production and distribution have already begun shipping in bulk and selling in larger containers or bottling in their markets to reduce their carbon impacts. Australian brands, for example, boast that carbon footprints to Britain are smaller than Bordeaux's.

On a more fundamental basis, the INAO has long advocated reducing vineyard environmental impacts with *lutte raisonée* practices. The approach uses monitoring of pests and vineyard conditions for more precise, and less frequent and smaller applications of chemicals. The CIVB and *syndicats* support that position and sponsored complimentary research and the dissemination of technical and regulatory information. Many producers go further with organic and bio-dynamic winegrowing and have been supported by organizations like CIVAM, but remain a small proportion. They face the same difficulty as organic producers everywhere in that they carry the ironic burden of paying for certification and a label, while those burdening the environment don't have to expose their practices on their labels. The number practicing defacto organic production, or close to it, is increasing because of the strengthening confluence of sustainable practices with the ideology of *terroir*. Most chateaux equate higher quality and expression of *terroir* with less chemical and machinery use, and hence there is return to labour-intensive practices.

The territorial enterprise also depends on providing for the health and pensions of its workforce, particularly important in an industry based on small companies without deep resources for either employer or employee. Fortunately France ensures that all workers, even seasonal, receive a high standard of benefits. The national system is expressed regionally through the MSA (*Mutualité Sociale Agricole* or Agricultural Mutual Society). Member payments and national subsidies spread out the costs of the health and accident insurance, pensions, and maternity costs faced by the self-employed, employees and employers. Although the chateaux proprietors believe this collective organization is necessary and beneficial for the welfare of their employees and the viability of their businesses, they also believe they give a lot more than they receive. In particular, they find their pensions meager when succession taxes burden them with great costs to hand down their businesses, and retirement funding suffers. It is fortunate that the state and region support what could be overwhelming costs because none of the organizations such as the *syndicats*, the CIVB, or the *Vigneron Independants* supplemented these welfare systems.

# Chapter 4
# Bordeaux's Territories:
# Leader and Aspirant

The Bordeaux brand draws recognition around the globe, but the territories that are the parts that make the whole. They construct the quality behind the reputation and define the many styles that give the region its complexity. The territories are the locus of direct self-governance by the winegrowers over their appellations – what standards will define style and quality, and how those standards will be enforced. Yet while the *syndicats* are masters of their territory's and appellations' fates they are beholden to the national government and the INAO for the mandate to govern their members, for permission to change their regulations, and are required to follow system-wide procedures. Although the latter requirement is often emphasized, it is the capacity to make use of the autonomy available to territories that allows some to distinguish themselves and demonstrate managerial and marketing innovations.

Bordeaux offers a spectrum of territories, in terms of size and renown, and in terms of collective organization and ambition. In this chapter we review the diversity of territories according to these dimensions, looking at common practices in governance and management, and compare two territories to emphasize the potential for effective self-governance. Saint-Émilion is offered as the exemplary of an innovative and ambitious territory that has capitalized on a long history of collective action and a compact geography. Blaye is described, not as a foil, but an explication of how winegrowers can overcome a fragmented past and space to establish a promising platform for the transformation from bulk to estate-winegrowing.

## The Territories

Distinct territorialization within the Bordeaux complex occurred concomitant to the initiation of a regional system in the medieval era described in the Bordeaux chapter. The Privalège de Bordeaux gave advantages to immediately adjacent areas and soon an area east of Entre-deux Mers developed as it was allowed to export its wines a month earlier than the high country. With an independent heritage, Saint-Émilion exported along the same trade routes and into some of the same markets as Bordeaux, and with its own traders and identity. Territorial differentiation increased in the Age of Reason as foreign traders, sought specific products meeting the tastes of their markets. The Dutch taste for sweet white wines stimulated the development of Sauternes and after they drained the marshes of the Médoc it later became home to the practices cultivated at Haut Brion and to cabernet sauvignon. Preferring the deep red style of Cahors, fertile river plains known as "palus" were also planted for the Dutch, notably south of Saint-Émilion. In later years these areas would be dismissed as poor *terroir*. The English trade, however, developed a taste for the "new claret," that developed in the Médoc. The Germans, Irish, and Scandinavians brought their influences, while wars, plagues, and weather further shaped trading patterns and shifted the pieces of the territorial mosaic.

In the 19th century national and international laws recognizing the origin of wines stimulated territorialization through a rebirth and expansion of winegrower organizations. Organizers of agricultural shows (*comice*) and national agricultural associations established local *syndicats*,[1] generating enough demand for a national law authorizing *syndicats* in 1884. The organizations diffused agricultural techniques, provided mutual assistance, especially in response to the repeated attacks of vine diseases, and lent some balance to merchant power.[2] The introduction of the railway, increasing urbanization, and the expansion of overseas markets encouraged greater planting and specialization by area. The odium, mildew and phylloxera diseases also caused local specialization. Knowledge was spreading and areas were becoming more focused on specific grape varieties. All of these trends led to greater territorialization.

The establishment of the AOC system was crucial to the development of the territories. Gestation was difficult, however. Prewar inititation of boundaries was followed by a 1919 law that allowed any winegrower to use an appellation after making a presentation to the courts. Competing applications proliferated throughout Bordeaux, and even among neighbours who were using the same grapes on very similar *terroir*. Some applications strove to make a chateau brand into an appellation, while others attempted to capture the reputation of an adjacent territory. The courts were overwhelmed and the high transaction costs imposed on both winegrowers and society impeded winegrower organization. In 1927, Capus, who had been on the pre-war commission, attempted his first reform to ensure an appellation of origin can only apply to products from the specific place named and to conform to "loyally and constantly" used practices.

The term "loyally and constantly" invokes winemaking traditions specific to a place and capable of demarcation by boundaries. However, some places had a good deal more history and unity to call on, while others had to invent these traditions or spatially expand their use. The variety of grapes used is resonant of the issue. Even the Médoc and Saint-Émilion only defined their grape varieties within the 100 years before the initiation of the AOC. Most other territories used a dozen or so varieties. In all of these cases creating a territory depended on the use and invention of history and social capital.[3] The territories that succeeded then, retain their first mover advantages today.

The laws and territorialization of the 1920s, however, only provided for horizontal differentiation, and not vertical differentiation,[4] insufficient for Capus' intention of making appellations representative of "grand vins." The 1935 *appellations d'origine contrôlées* attempted to deal with this problem by stipulating that *syndicats* define grape varieties, and control alcohol content, yield limits, viticulture and viniculture techniques and so on.[5] Recognized types of soil and sub-soil had to be defined. Collective self-governance was achieved by necessitating that appellations had to be established by a *syndicat* and they were responsible for determining the appellation's regulations. The creation of the Comité National des Appellations d'Origine oversaw the development of territorial regulations, but no effective system enforced the regulations. Still, the system dramatically reduced transaction costs for both society and winegrowers as testified to the ascension of Bordeaux's territories to appellation status within two years, with minimal government support, and with little recourse to the judiciary.

The initial AOC setup was not perfect. It lacked the resources and mechanisms to achieve quality control, and these and other flaws would have to be dealt over the years. Importantly, however, the system established the basis for self-governance. Innovations and emulations followed, enabling territories to differentiate themselves within Bordeaux and globally. The *syndicats* would devise not only, quality control systems, but also R&D programs, marketing strategies, and perhaps most importantly mechanisms to balance the different capacities, ambitions, and conflicts among the winegrowers. On the other hand, rigidities were built into the system in the crucial period of the AOC's initiation. In particular, the hierarchic appellation system, while reflecting the historical performance of territories up to that date and the accepted wisdom of *terroir* differences in the region, imposed that order on subsequent generations by writing different levels of performance into territorial regulations. For example, the Médoc *syndicats* decided upon much lower yields and tighter vine spacings than did the winegrowers of the broad Bordeaux territory or Entre-Deux-Mers or of Blaye. Thus in addition to challenging the perceptions of merchants and consumers in regard to the value of their wines, *syndicats* and winegrowers also have to deal with regulations that lower the common denominator and which have proven difficult to change.

Yet, facilitating change – upgrading the quality of the appellation's wines, introducing classification or other systems of internal differentiation, encouraging participation in territorial tourism programmes – has become the preoccupation of the *syndicats* as they strive to improve the competitive advantages of territory and winegrowers. The challenges to achieving such change vary with the characteristics of Bordeaux's territories.

The *syndicat* for Bordeaux and Bordeaux Superieur governs a territory that spans all Bordeaux, 55 percent of the wine produced (by volume), a similar percentage of hectares, and answers to 6,700 members. Nor does it simply represent two eponymous red wine appellations, eight dry, sweet, sparkling and rosé wines also use the name. Although dispersed throughout Bordeaux, most members are found in a 40-kilometre swath running east from the Dordogne and between Libourne and Blaye (broken by the Saint Émilion complex), the far south of the Gironde, in Sainte-Foy-Bordeaux and in Entre-Deux-Mers. The last two areas have their own *syndicats* and appellations, and similar mixes between territorial appellations and the Bordeaux and Bordeaux Superieur appellations are found in all territories. Indeed, anyone in Bordeaux can apply for these appellations and consequently it is often used as a last resort when winegrowers feel they won't get their wines certified elsewhere. On the other hand, many winegrowers don't bother with the appellations of their own territories because of weak reputations. Another key factor is that while the majority of production is blended into merchant or cooperative brands, it is the many ambitious estate winegrowers who lend credence to the territory's claim to fine wine *terroir*.

At the other end of the spatial and value scale, the Médoc comprises only 7 percent of Bordeaux's vineyards, yet it is characterized by a disproportionate degree of official differentiation. The two broader territories of the Médoc and Haut Médoc (closer to Bordeaux) include the prestigious communal territories of Paulliac, St. Estephe, Margaux, and St. Julien, and the lesser renowned Listrac-Médoc and Moulis. The communes range in size from Paulliac's 633 hectares and 53 winegrowers to St. Estephe's 1,244 ha and 135 winegrowers (Margaux is 1,413 ha, but 74 winegrowers). The high returns to winegrowing in these communes allow for an average property size of 14.5 ha, but many estates are much larger. Although the majority of winegrowers are cooperative members, the production from nine cooperatives accounts for less than 10 percent of the communes' total. In addition to territorial appellations are five levels of *grand cru classé*, the *cru bourgeois*, and the *cru artisan*. Despite this differentiation the Médoc is a red wine area emphasizing cabernet sauvignon with a requisite blending of merlot and cabernet franc (and allowing for carmenère, petit verdot and malbec). The *syndicats* are drawn together through the *Conseil des Vins du Médoc*, while the chateaux of the classifications have their own associations.

The Graves is officially of the Médoc and Graves family, linked by the gravel of their *terroir*, but it has both a longer history and offers a variety of red and white wines. It also has a classification system, but is notable for the desire of the winegrowers in Pessac-Léognan to distinguish themselves from the rest of the Graves by creating a new territory and appellation in 1987. The Libournais area includes the renowned appellations of Pomerol and Saint-Émilion, the so-called Saint-Émilion satellites and Lalande de Pomerol, plus Fronsac and Canon Fronsac. The top Libourne chateaux have achieved the same level of distinction as the Médoc grand cru and differentiate themselves with both territorial appellations and classifications. Virtually all the wine from the region is red, but the Bordeaux blend is inverted to a merlot-dominated wine balanced with measures of cabernet sauvignon and franc.

The côtes are six territories grouped together because their reputation derives from vineyards located on the right bank (northern) hillsides of the Garonne and Dordogne (except, just for whimsy, the Graves-de-Vayres on the left side of the Dordorgne, and which isn't included in the association of the côtes). The territories stretch someway in land from the river, giving them a history linked to the bulk white production that dominated Bordeaux until recently and which has left their *syndicats* with a difficult legacy to manage. The semi-sweet and sweet wines are clustered well south of Bordeaux on the Garonne. Cerons, Barsac, and the renowned Sauternes are carved out of Graves on the left bank, while Haut-Benauge, Sainte-Croix-du-Mont, and Saint-Macaire are carved out of Entre-Deux-Mers on the right bank. The cluster is united by the use of sémillon, sauvignon blanc and muscadelle grapes that the botrytis fungus transforms into sweet wines. Sauternes and Barsac are united through a classification system dating to 1855 that elevates Chateau D'Yquem far above its peers.

## Saint-Émilion and Blaye

Saint-Émilion and Blaye reveal much about how territory self-governance can provide winegrowers with competitive advantages within global competition. The two territories share a limestone escarpment that produces their most renowned wines and a winegrowing lineage that dates to the Romans. Yet, their fortunes have differed dramatically. Saint-Émilion is a compact community that has enjoyed close to two millenniums of consistent development of its winegrowing traditions. It uses that history to great effect, but moreover the territory provides a record of sophisticated collective action that arguably stands Saint-Émilion as the most innovative among French wine territories. Blaye's background is more mixed spatially and historically. Over the last few decades its winegrowers worked to overcome these disadvantages to establish a reputation in an increasingly crowded marketplace.

Democracy in each territory is similar, but reflects Saint-Émilions more complex evolution. Each *syndicat* elects a council with a representation ratio of about 1:30, an executive of several persons and a president. Winegrowers directly vote on issues such as appellation regulations. Across the AOC system, however, representation varies with the volition of winegrowers. Saint-Émilion and Blaye have increased equality by giving each coop member a membership in the *syndicat* instead of the legally permitted single voice. Yet, Saint-Émilion has gone on to recognize the different interests of winegrowers in the classifications and appellations, and the cooperative by reserving them council and executive positions. It also reserves internships on the council for young winegrowers and considered giving larger estates more influence.

The following comparison thus is not one of foils. Indeed within recent history both have accomplished the remarkable feat of converting from bulk wine to estate wine production (Table 4.1). Rather, the comparison shows some of the obstacles, advantages, and different paths that can be taken by territories within the same region and AOC system. It shows the extent to which volition and creativity in collective action is decisive for the reputation of the territory and for chateaux differentiation. The following sections explain how Blaye and Saint-Émilion compare on key points of territorial governance and how they fit in or stand out within the AOC system.

**Table 4.1    Saint-Émilion and Blaye Compared**

| | Saint-Émilion | Blaye |
|---|---|---|
| Number of Estates in Territory | 820[1] | 1,039[2] |
| Number of Proprietors in Territorial | 626[2] | 800[3] |
| Appellation | 99%[1] | 100%[3] |
| Proportion of Proprietors in Territorial | 406 | 400[3] |
| Association | 220[4] | 400[3] |
| Chateau Proprietors | 70% (83%)[7] | 46% (61%)[7] |
| Cooperative Members | | |
| Chateau Bottling | | |
| Average Size of Estate | 6.6 hectares[1] | 9.7 ha[5] |
| Vineyard in Territorial Appellations | 5,532 hectares[6] | 6,288[6] |
| Vineyard in Bordeaux and Other Appellations | 445 hectares[2] (8%) | 2,380[6] (38%) |
| Territorial Appellations | Saint-Émilion<br>Saint-Émilion Grand Cru | Premier Cotes de Blaye (red & white)<br>Cotes de Blaye (white)<br>Blaye (red & white) |
| Size of Region | 7,434 ha<br>(9 communes) | 62,376 ha (3 cantons + 1 commune) |

1   *Syndicat* viticole de Saint-Émilion.

2   Agreste 2001. *Recensement agricole 2000-viticulture (Agricultural Census 2000-viticulture)*. Direction Départmentale de l'Agriculture et de la forêt Gironde.

3   *Syndicat* des Premières Côtes de Blaye.

4   Brun Boidron 2004 *Bordeaux and its Wines* (Bordeaux: Éditions Féret).

5   Based 800 winegrowers in the allowed Blaye AOC area of 7,750 hectares, but average could be smaller because some proprietors own more than 1 estate.

6   Conseil Interprofesssionnel du Vin de Bordeaux (CIVB) 2003. *Marché des Vins de Bordeaux*: Campagne 2002–3 (Bordeaux: CIVB.)

7   ibid CIVB 2003 pgs. 33 and 50; bulk/bottling for merchants sales registered with the CIVB are subtracted from total sales (include bottled sales for merchants registered with CIVB).

## Legacies

Saint-Émilion can show visitors where the Roman poet Ausonius had trenches carved out of the limestone for the planting of vines. Formal territorialization, however, began in the Middle Ages when England gave the local burghers, the "*Jurade*," governance over eight parishes and their wine production. It "oversaw the production of fine wines, kept the iron that branded every single barrel, announced the start of harvesting, combated fraud, and abusive practices and destroyed wine that was judged unworthy of the name."[6] The *Jurade* imposed a general quality control, but differentiation by estate didn't arrive until the 18th century.[7] Inspired by science and economics, pioneers choose appropriate soils, drained fields, and selected the grapes characterizing Saint-Émilion. They tied quality to *terroir*, developed the notion of the *cru*, and lay the basis for chateaux. It was these differentiating proprietors that revived the *Jurade* after its elimination by the Revolution.

At the 1867 Universal Exhibition 37 proprietors won a collective gold and at the 1889 event 60 took a *grand prix*. They joined forces to overcome the prejudice that only grand estates could produce great wines with grand estates. In so doing they were taking advantage of Saint-Émilion's outstanding characteristic – it has been dominated by small properties since the release of feudalism.[8] A few landowners built stately homes and drew incomes from law, the military, or trade, but most lived and worked on their land. Small size drove differentiation, by *terroir*, by wine style, and by marketing. From the late 19th century, winegrowers adopted the term Chateau from the Médoc,[9] developed direct distribution channels, and enjoyed interdependence with merchants in Libourne.

When Parliament enabled farmers to form unions in 1884, Saint-Émilion established the first *syndicat*. Although immediately used to combat phylloxera, its importance grew as the appellation system took shape and facilitated formal territorialization.[10] Initially, a *syndicat* representing the Saint-Émilion commune wanted exclusive use of the name, but a rival from the seven communes of the ancient jurisdiction fought for equal rights in court. The *syndicats* merged, extending of the use of the name from 3,500 ha to a potential of about 5,500 ha.[11] Neighbouring communes, the Saint-Émilion satellites, were excluded but allowed to prefix Saint-Émilion to their appellation.[12] When the structure of the AOC system was defined in the early 1930s, not only did Saint-Émilion play a role in establishing vertical quality regulations, it also secured a critical basis for its own later innovation – all land in Saint-Émilion was designated as AOC, without distinctions by parcel or *terroir*. Saint-Émilion, furthermore, limited itself to a small set of red grape varieties. And although during the difficult pre-WWII period the *syndicat* focused on collective reputation, differentiation remained in view. The *syndicat* of "fine growths" had to be convinced a large-scale approach wouldn't cause their demise before they would allow the establishment of Bordeaux's first cooperative.[13]

Begat from the compromises of resource exclusion, the Saint-Émilion territory encompasses a diversity commonly grouped into four types of *terroir* with different fine wine potential. The limestone escarpment provides its backbone. It stretches several hundred metres west and a kilometre east from the town, covered with limestone-clay or red and brown clays. The majority of Saint-Émilion's most reputed chateaux are concentrated in this, the core of the Saint-Émilion commune. In the northwest of this commune is a broad flat plateau overlain by alluvial soils. The chateaux are somewhat less renowned except on the gravel knolls that produce Figeac and Cheval Blanc. Extending east through several communes, the third area comprises scarp slopes, their indentations and foot of the slopes. A southern aspect dominates, but different soils and subsoils produce a range of wine qualities. South and east to the Dordogne River extends a broad band of shallow sandy soils, superseded by a thinner band of sand and gravel by the river. This plain makes up two-fifths of the territory and produces its generic wine. The result of several hundred years of territory building was a diversity of *terroir* that required both a coherent image and a means for chateaux to differentiate themselves.

The Romans were impressed with Blaye's winegrowing potential as well and exports from the vineyards they planted continued through the Middle Ages. That progression was interrupted by construction and provisioning of a citadel in the 17th century. The "fury of planting" that swept Bordeaux in the 18th century brought new plantings on the escarpment and the birth of a reputation. By the 19th century the wines were classified by the Bordeaux trade and several chateaux exported to customers in Holland, Germany and Paris.[14] Chateaux rivalled those in Médoc in grandeur and used similar winegrowing techniques. Like Saint-Émilion, however, smallholdings produced by feudalism and post-revolution fragmentation dominated the landscape.[15] The winegrowers produced red wine in monocultures, with techniques similar to large chateaux, and for export, but because of their scale, transportation and barrel costs, Bordeaux merchants dealt with the distribution.[16]

A tiny port, Blaye is the focus of the territory and a limestone escarpment that stretches a few kilometres north and south along the left bank of the Gironde estuary. The vineyards concentrated about the edge of the escarpment and intermittently for about eight kilometres down its dip gave the territory its reputation and signature appellation, *Première Côtes de Blaye*. Farther away from the Gironde, in a rectangular area extending 25 kilometres north of the town and 25 kilometres to the east, mixed farming covered a landscape comprised of 20 soil types.[17] Originally, white varieties were grown for domestic consumption, but the coming of rail prompted farmers to produce low-value high-volume white wines for mass markets. These vines were grown on relatively fertile land, in polyculture and in yields elevated by fertilization – conditions not associated with fine wines. In the north, white wine was produced for the cognac of Charente.

Unfortunately, the leading producers of the escarpment were victims of the over-production and fraud that incited the creation of the AOC system. Most gave up differentiation and sold to merchant-blenders and this form of production remained their lot for decades after the advent of the AOC. The formation of the Blaye territory and *syndicat* was further weakened by the North's preference for inclusion in Charentes. Even after they were rebuffed,[18] out of administrative convenience, the territory was pulled together as three of the Blaye *arrondissement's* cantons. It was a heterogeneous territory of 60,000 ha defined more by inclusivity than exclusivity. Tellingly, a fourth canton, Bourg, obtained its own appellation because of its independent history and unity. Opposing appellation proposals slowed the development of a *syndicat* until 1925,[19] which then supported the existing delimitation and in 1929 the courts upheld their viewpoint.[20]

Five appellations had to be created to accommodate disparities in the quality of varieties planted across this wide area and to distinguish among them. Furthermore the *Première Côtes* was made available for use, not only for the slopes facing the Gironde, but also for the interior. Conversely, the basic white appellation could be used everywhere. This flexibility has advantages and disadvantages, but Blaye experienced swings in production of the two types of wine as a consequence and the winegrowers were left with a difficult legacy on which to build their territorial and chateau reputations.

## Chateaux Proliferation and Institutional Innovation

*Reputation Building through Collective Quality Control*

Although many Saint-Émilion winegrowers labelled themselves as chateau, until the war they still sold their wine by the barrel and trusted merchants to respect their differentiation. In the post-war period, viticulture mechanization, chemical treatments, education, cheaper equipment, plus increasing demand for fine wines, enabled the majority of winegrowers to vinify, age, bottle and marketed their own wines. As a result of this confluence of strategy and changing market conditions, the latent independence of Saint-Émilion's winegrowers manifested in the 1960s. The number of winegrowers calling themselves chateau climbed to 200 by 1980[21] and now the vast majority call themselves chateau or the equivalent. Seventy percent is bottled for direct distribution and another 13 percent bottled for merchants (Table 4.1). Saint-Émilion's cooperative remains important but even chateaux smaller than five ha bottle and market their own wine. The capacity for adding value is such that the average estate size is a mere 6.6 ha. In 1980–2000, driven by competitive differentiation, Saint-Émilion gained a reputation as one of the world's most innovative regions, in technologies, practices, organization and marketing, as epitomized by the "*garagistes.*"[22] These innovations were not limited to chateaux practices.

Saint-Émilion's pre-eminence in chateaux propagation was fostered by institutional innovation that set an example, but whose sophistication and ambitions have been hard to match. In 1951, Saint-Émilion initiated a wine tasting system that resulting in awarding a label granted to passing wines. This example was adopted and developed by Entre-deux-mers and several other territories. Saint-Émilion showed its commitment by refusing an appellation to the whole of the 1963 and 1965 harvests and only to 5 percent of the 1968.[23] In the 1960s the system, pushed by the CIVB, was adopted by the Bordeaux and Bordeaux Superieur *syndicat*. All AOC territories had to adopt it in 1972 when the INAO made it compulsory. But in Saint-Émilion, the potential for standardization and complacency inherent to the system was averted with interdependent appellations and classifications that fostered competition and recognized quality differences. Not unsurprisingly, the base of this system is the designation of *terroir*.

Maps of Bordeaux show territories coloured as contiguous spaces. Within those spaces, however, vineyards, and plots within vineyards, define land that can be used for appellation wine. Defining the geological aptitude of these plots is the primary technical duty of the INAO and the foundation of the AOC. What criteria the INAO should use for this definition, however, depend on negotiation with the *syndicat* and on their ambitions for their wines. Generally, *syndicats* are loath to

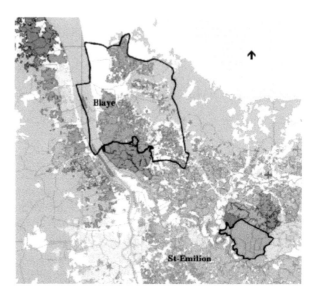

**Figure 4.1    Contrasting Definitions of *Terroir***

*Note*: Saint-Émilion's territorial *terroir* is contiguous, lacking any formal distinction among parcels. All other AOC land in Bordeaux, including Blaye, is mapped, bordered and classified at the parcel level. The AOC land is highly fragmented, although these distinctions rarely show up in the resolutions used in the maps published for most purposes.

change the AOC designations because removing it reduces a winegrower's income greatly and the attempt will be fraught with contention and legal aftermath. Bordeaux as a whole fell into this trap when growth in production occurred through the last decades of the 20th century. The conversion of non-AOC white wine vineyards to red AOC, was possible because AOC criteria were set in the 1930s, and Bordeaux, and the CIVB, couldn't impose reductions when the market became more demanding. Saint-Émilion has been an exception to this trap.

At first glance Saint-Émilion, renowned for its complex ranking of chateaux looks like a classic AOC territory. Its success, however, is based on the anomaly that status is not attributed to a geological definition of *terroir*. Except for the land given over to urban uses, all land in the nine communes is simply defined as Saint-Émilion, that is, all land is defined simply as Saint-Émilion AOC and no specific plots are given special designations. Thereafter, all rankings are based on the performance of the chateau. There are conventions and prejudices about which *terroir* is the best, but there is substantial room to gain a higher designation and added value from reputation. Saint-Émilion's success is all the more impressive because it was not an exception to the expansion of vineyards and volume that occurred throughout Bordeaux, nor to threats to quality.

The collective innovations began in 1951 when the *syndicat* created the forerunner of the AOC's tasting-based certification system. Then in 1954, it imposed the strictest yields in Bordeaux's red wine appellations and created a system of four appellations of Saint-Émilion, *Grand Cru*, *Grand Cru Classé*, and *Premier Grand Cru Classé*. Estate bottling was made compulsory in the last three of these appellations in 1972. In 1984, the system was simplified, placing all wines into either the Saint-Émilion and *Grand Cru* appellations, and turning the *Grand Cru Classé*, and *Premier Grand Cru Classé* appellations into classifications within the Grand Cru appellation. The ingenuity of the system is not its categorization, but the dynamism it fosters among the different categories while providing for long-term reputation.

The foundation of the system is the relationship between the two appellations. Like all appellations, both Saint-Émilion and *Grand Cru* have to be applied for each year, but winegrowers want to achieve the more respected and lucrative *Grand Cru* status. As a result they devote their best vines and grapes to the *Grand Cru* applications and learn how to cultivate their vines and make their wines according to those standards. Competition is maintained because the *Grand Cru* is restricted to about 60, although the proportion varies a few percent from vintage to vintage. The standards are consistently elevated, most notably by the pressures to reduce yields. The actual yields have been pushed closer to the basic yields of both appellations, resulting in a reduction of about 10 hl/ha over the last decade. The dynamic of competition is complemented by other requirements for the *Grand Cru*: a higher degree of alcohol, stringent testing by lot and label (Table 4.2). A critical innovation is a second testing in the year following harvest that provide a better indication of the wine's "aptitude for aging."

**Table 4.2    Regulations Governing the Saint-Émilion and Grand Cru Appellations**

| Appellation Criterion | Saint-Émilion | Saint-Émilion Grand Cru |
|---|---|---|
| Basic Yield<br>Ceiling Yield | 45 hl/ha<br>65 hl/ha | 40 hl/ha<br>60 hl/ha |
| Plantation Density | 5,000 vines/ha | 5,000 vines/ha |
| Natural Alcohol Content by Volume | $10.5 < x < 13$ | $11 < x < 13$ |
| Presentation of Wine for Testing | By appellation | By estate or label (marque) |
| Final Testing (if failing 1st and 2nd) | Samples taken of several lots (max. 300 hl) | Samples of several sub lots for any identified lot (max. 300 hl) |
| Aptitude for Aging | Not required, wine available for sale in year of harvest | Required, necessitating second testing for appellation in year following harvest |
| Chateau Bottling | Not necessary | Required |

Constantly raising the standards and maintaining inter-territorial competition cannot simply be imposed. The legitimacy of the testing process is constructed by transparency of arrangements, pre-examination tastings, appeals and publishing failure rates, and by fostering improvements. Three attempts at certification are allowed and failure rates are high, but the *syndicat*'s oenologist helps to select and improve wines to ensure almost all winegrowers produce some appellation wine. This rigour contrasts with critiques of *syndicat* certification rates that are based simply on final low failure rates. Extensive R&D, weather and pest monitoring support also raise standards. The *syndicat* has worked with the Bordeaux enology faculty for several years on monitoring and innovations. Thus where most appellations examine for a base standard, Saint-Émilion couples higher expectations with extensive support, thereby improving quality and legitimacy. Costs are kept low because winegrowers, merchants, and enologists do the certification voluntarily. A technical committee comprising 15 percent of all winegrowers guides these activities and hints to Saint-Émilion's true basis of ever increasing quality – it brings a significant number of the world's best winegrowers together in a place of constant competition and cooperation.

The QC system's worth to winegrowers is underscored by the fact that although AOC authority limits the amount of money that can be assessed from winegrowers for the certification process, Saint-Émilion has funded its more stringent controls through the dues of the voluntary membership to the *syndicat*. Membership to the *syndicat* from territorial winegrowers is about 99 percent. They were also among the first to obtain ISO 9001 certification for their QC system, enhancing value for the trade and consumers, while improving legitimacy to winegrowers. Winegrowers, merchants, and brokers testify that the *syndicat*'s use of the appellation and QC systems is responsible for the elevation of Saint-Émilion's quality and prices. Yet, the system doesn't work perfectly. Decision-making behind the proportion of *Grand Cru* to Saint-Émilion generic isn't transparent and one winegrower complained that supply should not be controlled by this mechanism. Hubert de Bouard, *syndicat* president for a decade until 2008, was aware of such concerns and wished to correct them, believing the system should constantly improve. Nor is the *syndicat* the only source of change. For several decades the cooperative has provided a different and important standard for grapegrowing and winemaking and the Chamber of Agriculture constantly upgrades its continuing education programs enabling winegrowers to keep abreast of change.

Improvements are not always easy to implement. Often when they achieve acceptance among winegrowers the territory's independent ways confront the INAO's demand for systemic consistency, particularly in regard to the geologic definition of *terroir*. In the late 1990s, when Saint-Émilion wanted to eliminate inferior land that from their AOC, it was pressured to simultaneously demarcate its classifications and appellations by *terroir*. Not wishing to lose the dynamism of its competitive system, Saint-Émilion aborted its designation of land. In other recent attempted reforms, Saint-Émilion winegrowers voted for a system to enforce vineyard regulations and to increase fees to pay for it and worked with the fraud department to develop chemical identification for each vat. The INAO didn't allow the former because increasing fees overstepped the legal bounds of the *syndicat* and the latter for fear of legal appeals.

Blaye's adaptions to demand shifts and regulatory changes during the depression resulted in the territory's transformation. Red vines were pulled, white planted, yields elevated, and vineyards enlarged. Eventually only three communes of the côtes retained the dominance of red and the red producers were forced to establish cooperatives to survive. The predominance of high volume, low quality white (much non-appellation) was embedded by the 1956 frost that incited replanting for immediate production with high yields.[24] That state remained until 1975, when Blaye began to respond to the shift in demand that was converting Bordeaux's vineyards to red wine production. Available statistics record that volumes of *Première Côtes* red went from an average of 27,553 hl/yr in the 1960s to 62,413 in the 1970s[25] while acreage grew from 3,106 ha in 1989 to 5,950 ha in 2003. Winegrowers using other Bordeaux appellations continually declined so that by 2003 total acreage of Blaye wines was 6,288 ha out of a territorial total of 9,688 ha. A significant portion of the remaining 3,000 ha converts to *Première Côtes* yearly.

Rebuilding quality was not simply a response to shift in demand. In the 1970s, when sales to merchants collapsed, winegrowers focused on their own reputations. Chateau bottling became popular, especially for tourists, and exports advanced from 6,000 hl in 1970–71 to 18,000 hl in 1978–9.[26] Winegrowers who withstood the lean years accumulated land and benefited from mechanization, while generational change and newcomers drove improvements. Today 46 percent of wine is produced under a proprietor's label and another 15 percent is bottled on chateau for merchants. In comparison with Saint-Émilion, Blaye estates require relatively larger economies of scale, with a 9.7 ha average and 156 estates are larger than 15 ha. Cooperatives also play a greater role for smaller producers. Five cooperative wineries serve the various areas of Blaye, and in 2005 these were amalgamated into one organization.

Blaye needed to manage the rush into its signature appellation from winegrowers whose land was given recognition at the establishment of the AOC. Their land varied in winegrowing aptitude, was often committed to inferior grape varieties, and vine spacings and other practices not associated with the production of fine wines. The moribund *syndicat* was awakened when quality control for all appellations became a legal requirement in 1974, and they used this authority and process as the principle means to counter the threat of massive vineyard conversions to the *Première Côtes* appellation.[27] In 1988 the *syndicat* purchased new headquarters to conduct certification in a scrupulous manner and ensure compliance with expectations for the appellation. Commentators began to note the managerial capacities of the *syndicat* and the promise of the territory.

Blayes ambitions became more evident when they reduced the land that could be designated AOC and to rewrite the appellation's quality standards. With the support and authority of the INAO, the *syndicat* reduced the territory's original 17,557 ha[28] to 10,000 ha by the 1980s,[29] and then to 7,750 in the early 1990s.[30] Blaye's reputation suffered and its place in the Bordeaux hierarchy was cemented because the quality standards set down in its regulations were indistinguishable from those of generic Bordeaux. These standards were raised to the level of Bordeaux Superieur and the other côtes in the 1990s, most notably changing the allowable grape yields. New certification facilities and processes were introduced and even a Saint-Émilion-style two-tier appellation system was considered. The latter was disallowed because the INAO considered it and the lack of *terroir* designation a Saint-Émilion exception.

Unfortunately, these steps didn't satisfy the majority of estate-bottlers who complain about free riding by underperformers. "They only see a bottle when they go to the hypermarket!" declared one interviewee. The majority are adamant that under-performers shouldn't be allowed to bring the appellation into disrepute and are responsible for their own fate. This conflict raises dilemma for the *syndicat*. Further removal of unsuitable appellation land is an option, but in the first decade of the millennia, was resisted as a divisive and expensive struggle at a time when winegrowers were looking for any income. The *syndicat* and individual members, therefore, pushed for improvement in the quality control system.

Although none considered the certification process perfect, Blaye winegrowers respected the impartiality and efforts of the *syndicat* members, merchants and brokers who volunteered for the taste testing. They did, however, want to eliminate the abuses of the sampling procedure, yet to avoid the cost of sampling and testing vat-by-vat. One member's proposal was very similar to the system of certification by chateau that is on offer in the new ODG, albeit without the use of third party intervention. How they will implement the ODG proposal for the needs of the territory is yet to be seen. The *syndicat's* request for a vin-de-pays outlet for some of the territories production has also been requited, thus enabling underperforming winegrowers to obtain some recognition for their efforts, while preventing the dissolution of the Blaye brand. Another method used to raise standards is to encourage underperformers to join a cooperative. The *syndicat* also provided a technician for advice before and after certification. It offers meetings for discussions of machinery, pest control, and other technical matters. Still because of its limited income the *syndicat* relies on the CIVB for technical newsletters and meteorological or pest notifications.

*Classifications and Reputations*

Classifications rank chateaux within a territory – generally, from premier cru classé status down a few levels. The forerunners were the 1855 classifications of the Médoc red wines (with Haut Brion an honourary inclusion) and Sauternes-Barsac sweet wines undertaken for the Paris Exhibition. These were done by the Chamber of Commerce at the behest of the French government.[31] Although the 1855 classification was a contemporary depiction of a ranking brokers had developed and modified over two centuries, both merchants and winegrowers grasped its broader marketing value and sought to institutionalize it. The classification became the property of the Chamber of Commerce, and they with the cooperation of the INAO and the winegrowers decided to fix the ranking of 1855 in perpetuity. In later decades other territories recognized the benefits of this system of differentiation and created their own formal classifications with the INAO's help.

Classifications are Bordeaux's most famous marketing tool, and also a very valuable means to legitimize participation in a territory. Within a territory, firms need to standout among the hundreds of their neighbours, to be recognized for the quality of their *terroir* and for their winemaking efforts. The classifications provide chateaux with this differentiation by means of a hierarchy of differentiation within the collective. Superior chateaux, called "locomotives," are intentionally designated for their ability to raise awareness for all. This mechanism works well enough that many wine writers claim that the majority of chateaux ride on the coat tails of the first growths. Chateaux competing for rankings also raise the overall level of quality in the territory and the classification is a reward that legitimizes their participation in collective efforts.

The creation of a classification is fraught with dangers, however, because it establishes disparities in status and income, some not necessarily reflected in the winemaking performance of the chateau. Unlike the Burgundian or idealized INAO system based on the geology of specific plots of ground,[32] Bordeaux's classifications are based on the chateau. This allows for recognition of performance if the winegrowers can collectively recognize these changes. In the case of the hugely successful classification of the Médoc, the INAO and the winegrowers chose to ossify the original standings for the rewards of long-term recognition, and disregard changes in performance and landholdings. The 1855 model was used in the establishment of the first Médoc Cru Bourgeois classification in 1932 and the Graves system in 1959. Winegrowers, critics and consumers, however, criticize this ossification, claiming it incites free riding by rewarding underperformers and under-rewarding over-performers.

Saint-Émilion created a classification where chateau performance is reviewed every 10 years and which became the model for the introduction of classifications and reform of older ones. Saint-Émilion's classifications are awarded on top of the Grand Cru certification. To be classified a wine's track record over the previous decade is examined. To avoid criticisms of vineyard amalgamation suffered by the Médoc, Saint-Émilion insists a chateau's wine must have consistently come from the same property during that time and used the same label. Thus consistency of performance is tied to consistency of origin, demanding proportionate investment in quality and providing the wine with a legitimate claim to *Classé* status. Other factors, consistent with *Classé* status, such as economic viability for maturing and bottling or vine age are considered. Marketing, reputation, and price are evaluated because the chateau is representing the territory. To provide impartiality the classification committee is made up of trade professionals proposed by the *syndicat* and selected by the INAO. Although this seems to be a system fraught with transaction costs, expenses are controlled by the appointees voluntarily devoting 18 months and dozens of daylong meetings.

As with the appellation dynamic, fairness is essential because promotions and demotions result in even larger differences in merchant support, wine prices, and property values. Not unexpectedly, these reviewed classifications are susceptible to legal reaction from winegrowers who when demoted face devaluation of product and property. The classification committee has always been criticized for its decisions, yet most demoted chateaux remain loyal. In 1985 Beau-Sejour Bécot's was demoted from Premier Grand Cru Classé because in the few years preceding the assessment it used grapes from newly bought adjacent plots. Claiming the land to be of similar quality the chateau fought the decision in the courts and had the sympathy of many experts.[33] The appeal was lost but, but the family have remained *syndicat* leaders. The family of Chateau Mauvezin has been in Saint-Émilion for 500 years, but despite demotion in 1985, the proprietor staunchly supports the *syndicat* and classification system. Grand Corbin Despagne was demoted from Grand Cru Classé in 1996, after the owner's prolonged illness, but his son attests to the objective fairness of the system. In 2006, Grand Corbin Despagne regained its status, only to experience the annulment of the classification when four demoted chataux launched a legal assault. They claimed that not all chateaux received the same level of scrutiny and that members of the committee had conflicts of interest. The courts decided with the plaintive, but the promoted chateaux, the *syndicat* and local politicians appealed to the Ministry of Agriculture and Senate for intervention. After a few appeals, in May 2009 they were awarded with a law reinstating the eight promoted chateaux, but requiring the *syndicat* to devise a new classification system by 2011.

The plaintifs against the classification recognized the pyrrhic nature of their victory as it left Saint-Émilion devoid of one of its most effective tools. Winegrowers realized that it would actually be those lower in the classification that would suffer the most and those who had been promoted saw years of work dismissed.

> "What is ironic is that those who complained in the first place will suffer most because they will become anonymous without it," said Matthieu Cuvelier, son of Clos Fourtet owner Philippe Cuvelier.

> Those involved with more prominent properties agree: "In fact this annulment will create a deeper gap between the well-known big brands and the lesser-known châteaux," a local wine broker told decanter.com. "Consumers will turn to the big names like Troplong Mondot, Angélus, Pavie Macquin and others."[34]

The consequences of submitting to external authority in the redesign of the classification could be several. The *syndicat* could lose control over selection of committee members. If their choice of wines is prioritized, the commitment of the chateaux to the promotion of the territory's reputation as reflected in their promotion and high prices could be dismissed. Certainly greater costs will be imposed to ensure the selection and time commitment of judges with no connection to any of the chateaux. For better or worse, the recourse to the courts compromises the ability of the *syndicat* to devise management tools with autonomy and affordability.

As Blaye evolved, several of its winegrowers sought an official means of differentiation. The territory also realized the need to identify its "locomotives." In 1990 12 winegrowers resurrected the cru bourgeois classification given to them by the brokers more than 100 years earlier.[35] The group was dismayed by the rush into the premier côtes appellation in the 1980s and the dilution of its reputation. Initial reactions within the *syndicat* were positive, but the self-imposed nature of the classification soon became divisive. One long-serving vice-president found himself ousted after becoming *cru bourgeois*. Today, those outside of the *cru bourgeois* deride it as an anachronism with no basis in *terroir* or vineyard or winery expertise. The *cru bourgeois* argue that they use their best wine, age with new barrels and do a thorough tasting test among themselves. They offer to open the classification to others, but so far everyone has turned down the offer. Complicating things, the Blaye group were sued by the Médoc *cru bourgeois*, who sought exclusive use of a title that they had to themselves for a century.

To counter the *cru bourgeois*, the *syndicat* transformed the unused appellation of Blaye into a symbol of the excellence. About 30 winegrowers were recruited to make wine under tougher regulations in small quantities with high quality and high prices. The *cru bourgeois* didn't argue against the initiative, but claim it will take a generation or two to develop a reputation. Further, the simplicity of the name Blaye brings little benefit when few in France recognize the name and it contradicts a system designating quality by listing qualifiers. Blaye can't bring them the marketing benefits the *cru bourgeois* brings now. Outside the *cru bourgeois*, people are also critical. Two recognized overachievers decried un-ambitious quality regulations and the attempt to represent the appellation through micro-bottlings rather than use leading chateau. Eric Bantegnies, of Haut Bertinerie, criticized how the appellation separates locomotives from wagons, citing damage done to Graves when Pessac-Leognan was decoupled.

The *syndicat* director claimed a real war had existed between the *cru bourgeois* and supporters of Blaye. But in a sense there is no conflict as the agenda of Blaye is to equal the reputation and prices of the Médoc communes, a status believed commensurate with the quality of the territory's wine. In tandem, the *syndicat* and INAO are taking control of the *cru bourgeois* classification in an effort extending up to the EU and when achieved, chateaux will be dropped and new ones added. This is dangerous because when the Médoc adopted a system of 10-year revisions, many demoted chateaux sued for reinstatement and resolution of that squabble is pending. Blaye's president, Christophe Terrigeol, wants to avoid these troubles, but will insist that only the chateaux consistently producing the best wine will be selected. The *syndicat* has a window to pursue this strategy because price differences amongst Blaye wines are not as high as among the Médoc chateaux.

The *syndicat* intended to promote the *cru bourgeois* heavily, but these plans have been complicated by the amalgamation of the Côtes into one appellation.

Winegrowers have discussed establishing a three-tiered classification for all of Bordeaux to reduce the conflicts rising from classification and to providing clarity to consumers. This proposal hasn't found traction given the linkage of classifications to territorial reputation and the complex competitive dynamic they provide. Several external observers simply question their utility. After all, wine critics, such as Parker or Peppercorn provide reasonably long-term evaluations of many chateaux. Even de Bouard of Saint-Émilion claimed the market selection to be more powerful. Yet, as seen by the intensity by which winegrowers strive to build classifications and to become classified, they justify the system with a longer view of benefits and with the knowledge that there are a lot of chateaux in Bordeaux. One winegrower put it thus "I wish a very long life to Robert Parker, but once he dies the chateaux that depend on him will have a difficult time."

*Classé* status almost assures a chateau assessment by several wine writers, whereas the unclassified chateaux have a more difficult time gathering this valuable attention. Moreover, while there is not a complete correlation of critic and *classé* rankings, the market consistently rewards the *classé* chateaux with higher valuations. Saint-Émilion's *Premier Grand Cru Classé* use their classification as the basis for marketing and information sharing. Within the Médoc there are similar associations for the Grand Cru, the Cru Bourgeois, and the Artisans, as there are for other classifications. Generally, these associations complement territorial efforts to provide estates with other opportunities to pursue differentiation. Occasionally, as with Blaye's Cru Bourgeois, by setting themselves apart they exacerbate tensions.

## Marketing Gown and Town

The promotional capacity of a territory is in the granting of an AOC,[36] but marketing, per se, was not an explicit mission given to *syndicats*. Rather, they were given the authority to "defend" their appellations. Nor is marketing supported in the ODG proposals. On the contrary, while *syndicats* or ODGs are authorized to extract fees from winegrowers for quality control they are forbidden from using the funds on marketing. Any collective marketing must be funded through dues that *syndicat* members impose upon themselves or other fund raising mechanisms. Here again Saint-Émilion displays a collective willingness to pay and participate as it provides a sophisticated marketing effort in concert with the dynamic of the appellations and classification. Marketing funding is agreed to by the membership, who notably extract much higher dues from the rankings of the *Grand Cru Classé* than from Saint-Émilion generic. Money can also be raised through other fund-raising efforts, such as festivals or taking advantage of government programmes. These funds are put to use promoting Saint-Émilion's heritage, drawing people to the region: to drink its wine, meet its winegrowers and to enjoy its landscape and history. As a compliment, ambassadors travel around the globe carrying the message of a highly differentiated territory.

The *syndicat's* efforts reach back to the 19th century when several chateaux in jointly participated in national and international competitions.[37] Its most evocative and efficient initiative came in 1948 when it resurrected the *Jurade* as a wine fraternity with a promotional purpose. Undoubtedly this resurrection was stimulated by the birth of the *confrérie de chavaliers de tastevin* in Burgundy in 1934, but Saint-Émilion can claim its *Jurade* to be the most ancient of the many existing conferéries. The invocation of the *Jurade* lends Saint-Émilion's quality control and fostering of a wine culture a depth of authenticity, and therefore has allowed it to become a ceremonial means to promote Saint-Émilion's traditions and to reinforce connections locally and globally. With the authenticity of clerical participation, the *Jurade* dress up in medieval robes and enact convivial rituals to celebrate the flowering of the vines, the opening of the harvest and other occasions. Positions in the *Jurade* are given to winegrowers and others within the local industry who make strong contributions to the *syndicat* or territorial reputation. Membership is also used to co-opt other important people, such as elected and administrative government officials, importers and exporters, journalists, artists and so on, who can help improve the prestige of the appellation. A special distinction is reserved for kings, queens, and other such romanticised anachronisms. The induction may occur in Saint-Émilion, at one of its several foreign branches, or may be used to give a foreign wine tasting a bit of pomp.

The town is of course an important tourist attraction in its own right. Purportedly named after an 8th century monk that initiated the growth of imposing monastaries and churches, Saint-Émilion is an admirably preserved medieval town that provides the territory an enviable tourist draw. Indeed, the town is given over to that purpose. Wine merchants, restaurants, tourist shops and confectioners selling the macaroon specialty line the streets. The main square congregates the monolithic church with the tourism office and the *syndicat's* wine shop (*maison du vin*). Located in the centre of the town, the *maison du vin* not only allows visitors to purchase the complete range of appellation wines, many from old vintages, but also provides an anchor of reference among the many merchant shops. The *syndicat* also runs a wine school that focuses on an appreciation of their wines, and other tasting opportunities abound.

More so than almost any other winegrower area, the identification of the political-economic area as a winegrowing territory is almost complete. This concurrence was recognized when UNESCO designated the original communes of the ancient jurisdiction of Saint-Émilion as the first wine region deserving to be a world heritage site. That this designation also included a portion of another commune the INAO added to the territory had no bearing. The buildings of the town of Saint-Émilion were given legal protection in 1986, but UNESCO's award recognized the remarkable viticultural landscape and hamlets of the rest of the territory. The *syndicat* played a significant role in achieving this designation, working hand-in-hand with the Saint-Émilion Tourism board, the town's government and the mayors of the other communes.[38] The territory is further integrated through the organization of events, fostering cooperation of the winegrowers with the tourist office and tourist companies, and supplying necessary information and material such as maps, guides, magazines, and chateau and wine route signage. These materials make the territory legible to tourists and buyers who meet a constant stream of events ranging from open doors at the chateaux, to the *Jurade's* seasonal pronouncements, to music and art exhibitions. The focus on tourism has not come without a cost. The town's population has declined to under 2,500, partially as a consequence tourism taking over. It has become very difficult for people who work in Saint-Émilion to live there or new buildings to be constructed. A sustainable development plan has been written to address these problems, but does so only for Saint-Émilion and doesn't integrate the other communes.[39]

On a more purely marketing perspective, the *syndicat*'s important activities are the wine tastings and events it arranges for the *en primeur* carnival. Each year the *syndicat* leads the reception of the wine world's leading critics and magazines, of restauranteurs, wine cellars, wine merchants and so on. Marketing beyond Saint-Émilion, however, is necessary as it sells over half of the value of its wine as exports and cooperates to reduce these costs. The CIVB organizes many of the tastings and other events that the *syndicat* and individual winegrowers may participate in. These are strategically focused on existing and developing markets in France, Europe and around the world. Usually, at CIVB organized events, such as Vinexpo or regional tours, Saint-Émilion participates in the Saint-Émilion-Pomerol-Fronsac group. This Libourne group also has the size to put together its own international and national events. When participating as a part of a group, however, Saint-Émilion is able to stage its own events, for example, inducting a merchant or artist from an importing country into the local chapter of the *Jurade*. Individual winegrowers participate in these events at their own expense, but the *syndicat* also serves a representative selection from the territory.

Saint-Émilion has also begun working more closely with the so-called Saint-Émilion satellites, despite the old conflict over the appellation name. This cuts costs and expands promotion for all, and is especially beneficial for the satellites, despite encountering some difficulties on tourism joint efforts. For that reason, the *syndicat* is leading an attempt to collaborate with other world heritage wine regions throughout the world (such as Cinque Terra, Tokay and Douro). In all these marketing activities, the general approach is to use the territory and its heritage as an umbrella reputation to support a diversity of *terroir*-based wines.

Where Saint-Émilion has its church and medieval town as a focus, Blaye has its own UNESCO recognized citadel, plus a port and cobbled market square with their extensive histories. Blaye's size makes the use of this centripetal force necessary. For example, despite offering the usual wine route signage, visitors face significant navigational and time constraints searching out winegrowers. After a few years the *syndicat* gave up on the open doors events that had become common in Bordeaux. The challenge to spatial organization is increased by the fact that the wine territory and political-economic definition of the area are not equivalent. The wine territory only contains three *cantons* of the Blaye *arrondissement* as the Côtes de Bourg exists as its own wine territory and canton. The town is close to 5,000 people and acts as a service centre for a more extensive and mixed farming community. Thus although regional government, with its loose administration of the arrondissement, the town and the tourism office supports the winegrowers, they are not solely concerned with its interests.

The *syndicat* has some misgivings about lack of coordination among local and regional authorities, but still works reasonably with them. Every spring it uses the citadel to match close to a hundred of its winegrowers with 10,000 visitors in a wine fair. A shop specializing in Blaye wines operates inside the citadel and the *syndicat* set up its *maison de vin* across the road. The *syndicat* piggybacks on citadel events such as the international equestrian competition. Its marathon, while touring the territory and running through several cellars, bands, etc., begins and ends in the citadel. Compliments are found in other local attractions, such as the asparagus fair's linkage between gastronomy and wine and the roman villa at Plessac.

Such initiatives cost money, and the Blaye winegrowers have imposed extra dues on themselves to support their marketing efforts. Because dues are linked to volume, Blaye's €1.5 budget is larger than most other *syndicats*. It is spent on trade fairs, entertaining merchants, journalists and critics, and organizing tastings. An example of the sophistication of the *syndicat's* efforts are campaigns it holds in Paris, St. Malo and Bordeaux, where a number of bistros and restaurants are each matched up with a few winegrowers who come and pour their wines. Complete membership in the *syndicat* by users of the *Premier Côtes* appellation demonstrates strong support for these initiatives. Still, care must be shown in choosing whose wines are to be selected for tastings and for shelf space at the *maison de vin*.

The core belief behind the *syndicat's* several marketing efforts is that small producers in Bordeaux lack the resources for promotion and that promoting the territory is necessary for success. Because of that awareness of the need for scale, and despite its relative size and citadel attraction, Blaye joined with the other Bordeaux Côtes territories to create a greater marketing presence. Blaye had been cooperating with the Premiere Côtes de Bordeaux, Côtes de Castillon, Côtes de Franc, and Côtes de Bourg since 1985 and in 1987 established a joint Connètablie (wine brotherhood) and in 2007 decided to extend their marketing association to a common label, unto which individual territories will be rendered in subscript. Côtes de Bourg dropped out of this association before it was ratified by the INAO in 2008. Designed for simplicity and to improve awareness, a further ambition of establishing a quality hierarchy or what to do with appellations such as Première Côtes de Blaye or of the Cru Bourgeois Classification have yet to be clarified.

**The Territorial Bottom Line**

The objective of self-governance is to devise territorial and chateaux reputations that will impress on competitive markets. In constructing self-governance, however, territories must negotiate diverse institutions of regulatory and supply chain governance that may help or hinder the capacity to control their reputation. Perhaps it is in this regard that the comparison between Saint-Émilion and Blaye is most instructive. Saint-Émilion has constructed its reputation, innovated in collective action, with a greater autonomy than not only Blaye, but other AOC territories in general. As the gatekeeper of *syndicat* authority, the INAO is obviously the most important external institution that the territories must engage with, but several others are also involved.

At the outset of the AOC system Saint-Émilion managed to establish its own system for differentiating *terroir*, or rather remained free from not only the rudimentary INAO designations of the time, and also from the increasing rigour that the INAO would impose. On the other hand, the INAO was critical to establishing and operating Saint-Émilion's appellation and classifications systems, and they still work together well on innovations and on the certification and other administrative matters. The two, for example, are trialling the use of *petit verdot* in the territory. Still there are tensions in regard to the *syndicat's* need for autonomy and the INAO's demand for systemic consistency, as in the confrontations over the redefinition of *terroir* and vineyard regulations. Blaye hasn't struck out on its own like Saint-Émilion, but has introduced considerable change with the INAO's blessings. The rejection of its dual appellations proposal, however, illustrates that not all initiatives are accepted. Perhaps as a result of that experience, Blaye works with the system and waits for the AOC system to create changes common to all territories, rather than develop a unique system for itself. It is not, however, timid about asking the INAO to make radical changes. Increasing the profile of the Côtes as a reputation, over the different territorial appellations exemplifies the ability to overcome the weight of traditions.

In the Anglo-American press, AOC regulations are usually attributed to some anonymous bureaucracy, painted as rigid and outdated, and an anachronism damaging to not only consumers, but also producer. The viewpoint often is set inside a story depicting the red-tape travails of British or American escapee to an estate in Bordeaux, or Champagne or Languedoc. In reality it is difficult to find a winegrower willing to disparage the INAO or the AOC. Despite occasional dissonance over regulation changes with the *syndicat* or even the imposition of controls on themselves, winegrowers respect the INAO for preserving territorial distinctions, for providing a reference for consumers, and for doing so professionally and with few resources. That is not to say that winegrowers won't criticize the shortcomings of the certification process or the occasional implacability of the "big machine." A few winegrowers, well aware of criticisms of the AOC and of the attention gathered by rebels, manage to rail against the AOC regulations to good marketing effect. However, especially in comparison to government impositions such as pension and health systems payments, labour regulations, customs, advertising regulations, and especially succession taxes, the benefits of the self-imposed AOC far outweigh its minor costs and inconveniences. Indeed, the AOCs provide the organizational base to fight against government overbearing.

The *syndicats* are effective agents at the commune, *arrondisement* and even regional level, but for dealings with the state and the EU the *syndicats* usually act through their federation or the CIVB. Saint-Émilion is generally content with the CIVB's interlocutions. Indeed, it sees the CIVB primarily as a regional coordinator and lobbyist in Paris and Brussels, and is less interested in its mediation of relations between winegrowers and merchants or its marketing. Blaye on the other hand remains bound to a contradictory relationship within the CIVB's distribution chain mediation.

Saint-Émilion does not have to rely on mediation of the distribution chain by the CIVB to any great extent for several reasons. The first is a co-evolution of merchants, often owning estates themselves, dedicated to selling Saint-Émilion and other wines from the Libourne area. A winegrower-merchant, whose influence spanned the 18th century, exemplifies this tradition: "Raymond Fontémoing knew how to handle his clientele. He would sometimes add two bags of chestnuts to the wine he sent his Breton correspondents, or a case of prounes or some hams. Naturally it was mostly his long-standing clients that received these gifts ..."[40] Unlike the Médoc wines, which were sold by Bordeaux merchants to upper-class British consumers, Saint-Émilion winegrowers sold their wines to wealthier clients in Northern France, particularly Brittany, Normandy and Paris, and also to Belgium and Holland, and also into the northern German principalities. In the early 20th century, entrepreneurs from the Limousin and Correze regions, such as Mouiex and Janouiex, further developed sales in Paris, Liege, Lille and Brussels. These northwestern French and Belgian markets remain strong today, particularly for mid-range wines.

Second, Saint-Émilion reduced its needs for the CIVB's mediation by strengthing chateau bottling and building a reputation that rivals Bordeaux. The vast majority of winegrowers can therefore use the Bordeaux marketplace and the more renowned chateaux exercise significant leverage with merchants. Winegrowers don't abuse this relationship even if they have loyal consumers, because a diversity of merchants are required for global distribution of very small volumes. Most of the top chateaux offer tours to improve their relationship with existing or potential customers, but will not sell directly. Third, many winegrowers have taken on merchant roles to sell their own wine, that of others in the territory, other Saint-Émilion and Bordeaux chateaux, and from other territories. Families such as the Ouzoulias do so to extend and make their brand lines more competitive. Fourth, the *syndicat* co-opts merchants into the certification and classification processes to give them ownership, and through the parameters of classification, penalizes firms which don't use the marketplace. Purists decry this influence on evaluation, but Saint-Émilion's diversity would unlikely survive without supporting its distribution system. In short, Saint-Émilion has devised its own system for distribution chain mediation.

Blaye remains bound to the CIVB's mediation of the distribution chain because of its volumes of bulk wine and because the relatively low value of its chateau bottled wine keeps it in competition with brand wines. Hence winegrowers expect leadership from the CIVB. Yet they also describe the CIVB as prejudiced to preserving Bordeaux's appellation hierarchy and to the merchants' interests. They also criticize ineffective and expensive advertising campaigns and claim the CIVB to be out of touch with realities in the vineyards. Although there are many differences among the Blaye winegrowers – estate versus bulk winegrowers, newcomers versus the deep-rooted – their solidarity was evident on these points.

The intense expectations toward the CIVB arise because Blaye hasn't achieved Saint-Émilion's territorial reputation – as symbolized by the fact that few winegrowers make effective use of the Bordeaux market to increase prices and have difficulty obtaining marketing efforts from the merchants. Substantial friction arises between the two sides, but merchants have neither capacity nor incentive to represent dozens of chateaux adequately, and are likely to free ride on reputation of chateau and appellation.[41] Consequently, most chateaux sell first and second wines direct to specialist stores, hotels, restaurants, and private customers. The lack of territorial market power resonates in bulk sales as well because, according to the winegrowers, the merchants refuse to offer prices reflecting costs or quality. A revealing difference is that where Saint-Émilion's merchants are rooted in their territory, Bordeaux's merchants don't even have offices or warehouses in Blaye. Blaye winegrowers band together in marketing groups, some with external winegrowers to overcome some of these limitations.

# Chapter 5
# Napa: *Terroir* to the New World

In an episode from TV's *Odd Couple*,[1] Felix, reeling from a money crunch, despairs that he will be reduced to drinking domestic wine. Today, few wine drinkers in the US or elsewhere would see that as unfortunate end. The winegrowers of Napa can take a good deal of the credit for that change in appreciation. Napa is the undisputed leader of fine wine production in the US and its reputation for quality is the foundation for the California wine industry's reputation even though Napa produces only 4 percent of the state's wine.

Napa's reputation eclipses that of any of its winegrowers, yet it is they who created this renown. Underlying the territorial reputation are hundreds of excellent wines, each blended from an appreciation of the valley's diverse *terroir* and creative viticulture and winemaking. Moreover, Napa's winegrowers have excelled in adapting and creating collective mechanisms that overcome weaknesses in the value chain and capacities for self-governance. Napa's collective experience is rooted in the 1860s, but continues to evolve with changing market and external governance forces. Perhaps the most outstanding thing about Napa is the extent to which it has had to go it alone, *sui generis* in marketing, value chain, appellation, social and legal initiatives.

## Diversity Denied

The early history of Napa is a cautionary tale. The development of the California wine industry, like counterparts elsewhere, was dominated by merchants. Based in San Francisco, they blended or otherwise modified grapes, must and wine bought from grapegrowers for sale in eastern markets. Initially production depended on the native mission grape, but towards the end of the 19th century the blends became characterized by inferior European varieties fortified with chemical colouring and later with zinfandel,[2] and often sold as "burgundy" or "chablis." These names would become the "semi-generics" of international trade negotiations. Little of this wine was labelled as originating in Napa or the other promising regions.[3] Yet among the 200 grapegrowers and winemakers who were drawn to Napa by the late 19th century, several were committed to producing fine wines. Europeans, such as Jacob Shram, Henry Pellet, and Charles Krug brought knowledge of winegrowing and higher expectations, and initiated the cultivation of the slopes. Americans, George Crane[4] and Hiram Crabb held similar expectations and invested in vine varieties, rootstock and vineyards.[5] The Finnish sea captain, Gustave Niebaum, at Inglenook, created not only an estate winery, matching grapes to *terroir* and

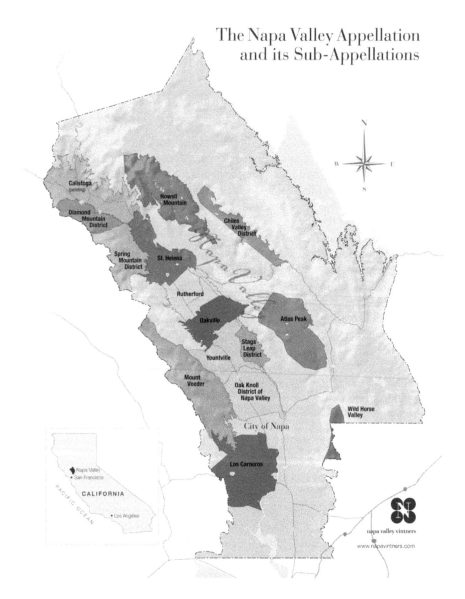

**Figure 5.1    Map of Napa Valley and its Appellations**
*Source*: Napa Valley Vintners.

vinifying and ageing in separate vats, but also pioneered estate bottling and promotion.[6] These practices were rare even in Bordeaux.

Napa's wines began to outshine other regions in competitions and premium sales by the late 1870s,[7] but not because Santa Clara or Sonoma were less gifted with *terroir*. The entrepreneurs who built Napa's reputation were conscious of their interdependence. They joined the 1859 formation of the Sonoma-Horticultural Society[8] and then the Grape Growers Association of Sonoma, Napa and Solano Counties in 1871. Members shared information on soil cultivation, pruning, vine selection and grafting, and formed a political voice.[9] Later Krug, Pellet and Ewer founded the St. Helena Vinicultural Club in 1875 and membership rose to over 100 in a few years.[10] A collective effort against phyloxerra expanded into diffusing techniques, working to lower freight rates, developing eastern markets and wine tastings. They also imposed quality controls, pledging not to sell grapes to wineries that used sugar during fermentation.[11] Collective efforts spread north with the founding of the Napa Valley Viticultural Society in 1881[12] and the St. Helena cooperative in 1988. Several winegrowers worked together to develop marketing channels to bypass San Francisco wholesalers.[13]

Remarkably reminiscent of Saint-Émilion, Napa united to provide samples for the 1989 Paris Exhibition,[14] and won 20 of 34 awards given to American wine. Although the vinicultural association provided a forum, a hallmark of Napa's collective action was, and is, informality. Technical information was shared freely, precedents set, and expectations raised. George Crane, for example, used newspaper articles and letters to compel winegrowers to improve quality and resist the corporate temptation to homogenize,[15] while others applied a "moral lever" to invoke a community approach to the improvement of quality.[16] It became a tradition for owners to help their winemakers when they moved on or set up on their own.[17] The newspaper published accounts of association meetings, followed the industry and the yearly output of producers.[18]

Napa engaged state or national organizations with mixed effectiveness. Berkeley and Eugene Hilgard taught fermentation, vine selection and location and Hilgard's advocacy of quality over quantity resonated with the estate winegrowers. The State Vinicultural Society and agricultural fairs were used for promotion and governments were routinely lobbied. In 1878–9 Napa joined other Northern California producers to protest against allowing the importation of French wine tax-free.[19] It had less happy relations with the Agricultural Department's State Viticultural Commission (est. 1880). Despite the benefits of its technical literature and regional commissioners[20] and even though it established a viticultural station in the Valley, Napa winegrowers were unmoved by the commission's work. They supported Hilgard in a battle with the commission over legitimacy and funding. To the antagonism of other regions, Napa elicited the abolishment of the Viticultural Commission in 1895 for failure to deal with local concerns of over-planting and pyhloxera resistant rootstock.[21]

The evolution of diversity and cooperation in Napa coincided with a broader boom driven by economic growth and phyloxerra's devastation of European vineyards. Vineyard acreage increased from 3,635 in 1879 to 20,763 in 1890, while grape tonnage increased from 11,600 in 1879 to 51,000 in 1888,[22] levels not seen again until the 1970s. California acreage increased from 36,000 to 90,228 in 1880–90 and production from 4,800,000 to 19,900,000 gallons.[23] When

Europe recovered, however, California found itself with excess production and in a depression. Phyloxerra then took its turn at California's vineyards. Price-cutting, consolidation of the industry, and rapid decline of estate wineries ensued. The California Wine Association (CWA) amalgamated seven San Francisco merchant-winegrowers[24] and by 1902 controlled over 50 wineries and 30 million of California's 40 million gallons of wine. The CWA integrated production and marketing and centralized activities in San Francisco. "The wines were stored, blended to a uniform standard, bottled, and then shipped for sale under the Calwa Brand, with its trademark of a young Bacchus, accompanied by the California bear, standing at the prow of a ship whose sail bore the seal of California. Thus the idea of a standard, unvarying product bearing a brand identity was introduced into the California trade."[25]

Napa, through Charles Carpy's Napa Valley Wine Company, played a minor role in the CWA and the St. Helena newspaper editorialized that the CWA offered a lifeline in difficult times. However, the consequences of blending wines into inferior, ordinary, superior, and fine grades soon became apparent as did the monopsonist's capacity to store and wait out producers and the market. The community realized that "the main purpose of the CWA was to crush the winegrowers."[26] Some producers helped to organize the California Wine Maker's Corporation (CWMA) as an opposition, but this challenge was crushed with the surplus production and price war of 1897. Independent producers fell from over 140 in 1889 to less than 40 in 1909. By prohibition the CWA controlled more than half the Valley's wine.[27] [28] The only compensation for Napa's reputation was the industry's awareness that the CWA's higher-end brands owed their quality to Napa grapes.[29] Yet when prohibition approached, organized resistance floundered.[30]

The California Grape Protection Association (CGPA) didn't disassociate themselves with demonized saloons and liquor because they depended on selling brandy and fortified wines and distribution by liquor companies. When the CPGA eventually abandoned that alliance and tried to get wine exempted, Napa withdrew on the idealistic basis that poor should not be driven to bootleggers while the rich enjoyed their wines.[31] When prohibition arrived, Napa suffered the same ironic fate as the rest of California's vineyards. Plantings expanded to supply home winemakers with transportation resilient but poor quality varieties. Napa's grapegrowers actually made record profits as its proportions of Alicante, Muscat, Zinfandel, and Carignane rose higher than state norms.[32]

At prohibition's repeal few quality vines survived, but the remnants were crucial to the rebirth of estates. The ownership of Beaulieu provided a bridge to French winemaking and after hiring the French trained Andre Tchelistcheff, stimulated the transformation of viticulture and winemaking.[33] Inglenook's uncompromising commitment to estate wines produced Napa's archtypical role model. Newcomers in the 1930s, Martini and Mondavi were soon winning competitions. A few pre-war wineries (Larkmead, Lombarda, Nichilini and Deer Park) survived by selling a combination of bulk and estate wines. In the 1940s and 1950s, Freemark Abbey, Souverain, Mayacamus, Mt. Diamond, Sutter Home, Stoney Hill, and Only One opened with the intention of making premium wines.[34] This small group revived the St. Helena Vintage festival and fostered external organizations like the Wine Institute.[35] During the war they initiated the Napa Valley Vintners Association (NVVA), which from convivial origins became the organizational force in the valley. With an effective use of Hollywood and Harvard the NVVA rebuilt Napa's reputation in the 1950s. Externally, this small group initiated the Premium Wine Producers of California and set out to foster an American wine culture. Tchelistcheff founded the Napa Valley Technical Group to generate and diffuse information among winemakers and interact with oenology at UC Davis.

Yet throughout these decades bulk wine production dominated. Two cooperatives accounted for about half of production and a dozen other bulk producers came into production by the 1940s.[36] By the late 1950s Gallo was buying 65 percent of Napa's production.[37] Few of Napa's winemakers of the pre- and post-war decades were estate winegrowers, rather it was a period emphasizing vertical quality; in Lapsley's terminology, they were "quality producers."[38] They planted or bought the better grapes that became known as varietals, modernized wineries, and developed their winemaking expertise.[39] Improvements in the latter two were emphasized, in the belief that quality was made in the winery. The grapegrower, Beckstoffer, pointedly called it the era of "magic chefs." Market demand for generic "burgundies," "chablis" or fortified wines forestalled estate winegrowers.

Despite a gradual evolution in technique, investment, and organization, by 1960 industry consolidation and prohibition reduced the number of wineries to a 20th-century low of 25. Things soon changed dramatically. A second wave of variety began to build in the 1960s and seems yet to crest.

In the 1960s, ghost wineries from Napa's first glory period were reinhabited, and true to the ethos of the era, often by corporate escapees. Robert Mondavi built the first new winery in decades and became an incubator for talents such as Grgich and Winarski. The US fine wine market grew, assigned high prices, and spurred on more winery development. The number of wineries grew to 51 by 1977, and then doubled to 110 in the next four years.[40] By 1990 there were over 200 wineries in the valley, by 2003, over 260. The first phase of expansion of independent wineries was neatly portrayed in the term "boutique" wineries, conveying their size and their handcrafted appeal. Increasingly, however, as Napa's reputation escalated, so did the value of its land and the difficultly for less moneyed people to pursue their dream. In the 1990s owning a vineyard in Napa Valley became the ultimate status symbol to Silicon Valley millionaires.[41] Still, doors remain open. Some people establish wineries relying on purchased grapes and may eventually buy vineyards; others buy vacation homes only to be drawn into the business; and some winemakers and grapegrowers have built reputations big enough to finance wineries or estates.

**The *Terroir* Transformation**

The 1976 "Judgement of Paris" is recognized as a turning point in Napa and US wine production. Irrespective of how that tasting of top French and US wines may be interpreted, the majority of American wines were from Napa and they took their place alongside renowned benchmarks for quality. Less considered is the fact that this event marked the triumph of *terroir*. The wines were not from Mondavi or Martini, but small producers making wine from specific vineyards. Even though some of these successful wineries were not estates, they sourced their grapes from one or a few vineyards. Stag's Leap Wine Cellars drew its grapes from its Stag's Leap vineyard and Heitz Cellars from Martha's Vineyard. These examples define the two dominant means of capturing value from *terroir*. Stag's Leap Wine Cellars bought and converted a prune orchard to secure grapes for its signature wine for the long term. Heitz bought his grapes from the grapegrowers Tom and Martha May who developed the archtypical vineyard designate.

The shift to *terroir*-based production took a few decades. In the late 1960s and 1970s, Brounstein, Forman, Heitz, and Winarski focused on expressing the characteristics of specific plots. They were assisted by grapegrowers, such as May, seeking to optimize their land. There was not a massive switch to estate winegrowing or vineyard designate wines. Technical, investment and psychological barriers stood in the way. The testimonials of many winemakers describe how their thinking was turned on its head by a trip to France or an internship in one of its vineyards. These encounters challenged some of the information Napa had gained in its otherwise beneficial relationship with Davis. The transformation hinged on the ownership of most land by grapegrowers or other landowners, rather than the winemakers. Land would have to be bought or relationships developed with the grapegrowers.

The conversion of the grapegrowers to sales to Napa wineries was completed by the 1980s. The co-ops closed down and Gallo went elsewhere as waves of new wineries opened up, each demanding higher quality grapes. At the same time vineyards expanded from 10,000 acres in 1960 to 45,275 in 2008. This conversion includes agricultural or fallow land and, especially on the mountainsides, freshly cleared land. Long-time Napa grapegrowers, other farmers, estate winegrowers, and external investors were responsible, as were purchasers of country homes and other real estate. Overall there was a large shift in the ownership of land and the proportion of land owned by grapegrowers and vintners respectively. According to Sullivan's estimates, in the 1960s wineries only owned 25 percent of vineyard acreage; by the 1970s wineries held 35 percent of vineyards; and by the early 1990s they held 45 percent.[42] A 2001 attempt to define ownership of the valley found over 1,000 vineyard owners, but that 70 entities own 25,163 acres or roughly 63 percent of the valley's vineyards.[43] Of that number, 55 predominately winemakers owned 44 percent and the 15 grapegrowers with holdings over 100 acres own 7,414 acres or 19 percent of the Valley.[44] The ownership of the remaining 14,833 acres or 37 percent is fragmented among some 300 wineries and around 800 grapegrower vineyard owners. Since 2001 consolidation has increased, notably Constellation acquiring Mondavi.

Vigorous development of estate winegrowing could be expected given European precedents and the need for winemakers to secure the origin of their grapes in order to build reputation. Robert Mondavi adopted this strategy and from minimal landholdings eventually bought more than 1,000 acres.[45] Others took longer, preferring the flexibility and lower upfront costs of purchasing. Phelps switched to avoid uncertainties such as being outbid for grapes, sale of vineyards, and grapegrowers choosing to make their own wine. Duckhorn's rationale includes the fact that estate wineries are valued more than those relying solely on a brand.

Vineyard designates gained popularity in the 1990s with the rise of cult wines and because the success of these limited bottlings depended on a fragment of land. The attention and prices paid to cult wines attracted larger wineries to buy vineyards for designates or to purchase grapes from them. Several grapegrowers use designates as the key to their value-added strategies, pushing winemakers to use them on the label. Creating vineyard designates, however, takes some time and effort. The characteristics of a vineyard have to be defined, grapes matched in variety and proportion, and because regulations require forsaking blending other grapes (albeit with 5 percent allowance). Winegrowers must consider what such selectivity will do to their product range and sales.

*Terroir* is of course intimately linked with grape selection. Until the *terroir* transformation, however, grape choice was driven by the association of quality wines with varietals, irrespective of correspondence with soil. Since, Napa has moved towards a set of grapes that provide both a match with *terroir* and a strong identity. When Nathan Fay planted cabernet-sauvignon in Stag's Leap in 1961, there were only a few hundred acres in Napa and only 800 in California. By the millenium, in image and reality, cabernet-sauvignon was king. By 2002, 35 percent of Napa was planted in cabernet, dominating from Calistoga in the North to Napa City in the South. Chardonnay and its climatic brother, pinot noir, retreated to cooler Carneros. The success of the leading cabernet pioneers led to extensive planting of the grape by many new entrants and heated competition. The belief of wine writers and consumers that cabernet-sauvignon produces the finest of wines contributed to this impulse. The dominance imposed itself without any territorial governance, but regulations did have an impact because. Any wine labelled as a varietal requires 75 percent of that grape in it, and if it claims an AVA designation, 85 percent of the grapes must come from that region. If a Napa winery was going to produce a cabernet, it would be with primarily with grapes from that region. Most important, however, was the recognition that the valley and Cabernet harmonized, since the founding of To Kalon and Inglenook, Napa's best wines were cabernets[46] and viticultural research vindicated the match.

Cabernet-sauvignon's climb to dominance did not go unchallenged. In the 1970s boom, chardonnay was the wine of choice and dominated plantings throughout the valley. But eventually, cabernet prevailed, suggesting that selection by *terroir* can balance the vicissitudes of market forces. No winegrower would deny paying attention to the varietals demanded by the market, but they also claimed that you can never chase the market as a popular trend could shift before investment can be recouped. They found the best strategy is to plant what they think the market will accept, but only based on the quality that can be created by the mix of grape and *terroir*. Quality and differentiation are the best bet for a smaller producer because large producers loose the ability to work with *terroir* when servicing the tastes of a broad market.

The *terroir* logic also works against complete dominance of cabernet-sauvignon in a variegated territory. Merlot, zinfandel, and sauvignon-blanc, are planted in substantial quantities, while others such as granache, nebbilio or tocai friulano exist in tiny quantities. Merlot exists in significant acreage because the majority of Napa's cabernet-sauvignon varietals are really a Bordeaux blend or meritage. Other Bordeaux grapes (cabernet franc, petite verdot, malbec) are grown for the same reason. While some plantings are remnants of an earlier era or of interest primarily to devotees (including of necessity, the winegrower), most require some experimentation with *terroir*. Duckhorn spent 25 years experimenting with locations to produce a merlot varietal. More extreme, Cosentino tests different grapes throughout the territory's valleys and grows or buys 15 different varieties. Out of preference for Rhone style wines, Joseph Phelps experimented with Napa's *terroirs* until he found a suitable match on the cooler mountain slopes and Carneros.[47] The high price of cabernet keeps a lid on experimentation with lesser valued varieties. Even America's grape, zinfandel, while occupying a similar climatic niche and designated as heritage vineyards cannot sell at cabernet's price and acreage is decreasing.

## Appellations and Brands

### Appellations and Sub-appellations

Long before *terroir*, Napa's winegrowers had constructed a collective reputation for producing superior wine. Yet when the opportunity presented itself, the winegrowers, sought the benefits of an appellation system that could provide legal protection to their reputation. The AVA system, however, was setup by bureaucrats, not winegrowers and it sits uneasily within an economic system unwelcoming to non-corporate designations of trademarks and territorial collective action. Thus in addition the contentious boundary setting traumas experienced when most territories are initiated, Napa has had to concern itself with continuing legal battles necessary to overcome the shortcomings of the AVA system. Perhaps more importantly creating appellations in Napa has established a platform for sustained territorial organization by grapegrowers and winemakers and for the evolution of constructive relations between them.

In the mid-1970s, the BATF realized it vague wine-labelling system was not conducive to the development of premium wines. Its preference, however, was for the states and industry to develop an appellation system. At that juncture Gallo and the Wine Institute sought to designate 14 counties as the North Coast wine region.[48] In reaction, Napa and other Northcoast winegrowers organized to demand the BATF retain control, and it responded with two years of negotiations over the constitution of the system. The NVV argued that appellations should be based on geographic and historical criteria decided by a committee of four vintners, two grapegrowers, one BATF official, an independent wine expert and a county farm advisor. The recently established NVGGA proposed an equal number of grapegrowers on the advisory board. The BATF declined both these suggestions to make its employees the decision-makers on AVA applications. The grapegrowers' claim a greater victory in achieving requirements for grape variety and origin proportions. Seventy-five percent of the nominal grape variety is required for a varietal label. More importantly 75 percent of grapes for a county appellation must come from that county, 85 percent of an AVA wine's grapes must be from that AVA, and a 95 percent proportion is required for a vineyard designate. Beckstoffer claims the leverage of California and American Farm Bureaus helped the Napa grapegrowers achieve this respect for the value of their land.[49]

Napa winegrowers were the first to apply for an AVA after the system started up. The historical, geological, cartographic and other aspects now formalized as a written application were then argued in an open forum of stakeholders. It was a contentious discussion because while the majority winegrowers wanted exclusive use of Napa's name, bringing wine or grapes from other regions to bottle as Napa was a longstanding practice, and by leading winemakers. The application turned on whether the boundaries should be defined by the contours of the Napa river watershed, the elevation of the vineyards, or the entire county. The most important question was whether the Pope and Wooden valleys should be included, that is, include some valleys not in the Napa river watershed. The grapegrowers were overwhelmingly in favour of just the watershed, and many vintners backed them in this.[50] Others, including Robert Mondavi who bought grapes from the Pope Valley wanted these valleys included.[51] In a less than transparent decision, the BATF allowed almost the entire county, including the disputed valleys to be included. Thus, the BATF set a precedent for a generous inclusiveness, for which it has often been criticized.[52] The decision has rankled many inside and outside of the valley since. Wine critic James Laube described it as an egregious mislabelling, equal to other abuses of the appellation, or for that matter, allowing 15 percent of wine to be brought in from other regions.[53] The impact would, however, be reduced by dividing Napa into sub-appellations.

The most knowledgeable person on appellations was silent at the BATF hearings. Andre Tchelistcheff brought concepts of area ecologies and microclimates from France and wouldn't endorse a broad appellation. He divided Napa into 16 different appellations.[54] With the evolution of the valley, he would see his vision realized. Carneros was the first and claims to be the first "American appellation granted based on micro-climatic influences and *terroir*."[55] A distinction reinforced by the appellation overlapping two counties and the larger Napa and Sonoma AVAs and requiring substantial effort to overcome collective and regulatory hurdles. But the application process generated long-term territoriality in the form of the Caneros Quality Alliance with its sophisticated marketing and research programmes. Other applications, such as those for Stag's Leap or Chiles Valley, have been more rancorous, involving battles over borders and requiring suppression of neighbourly animosity.

Since Carneros there have been 14 applications for AVAs. At the outset of this development the NVVA and NVGGA set up an appellation education committee wishing to bring some coherence to their creation.[56] The Robert Mondavi winery proposed organizing the sub-appellations around towns[57] and purported geological differences like the Rutherford bench and Oakville bench. Although criticized, these distinctions gave rise to some AVAs and most provided the basis for like-minded territorialization. The AVAs consist of a dozen or two wineries and a smaller number of grapegrowers, and offer more cohesion and differentiation than Bordeaux appellations. Indeed the smaller ones have fewer winegrowers than Saint-Émilion's Premier Grand Classé and none are the size of the Grand Cru Classé classification. All the same, many were concerned about the "balkanization" of the valley and reducing Napa's reputation. In 1990, after years of petitioning the State Legislature, the NVV obtained its law of conjunction, requiring those using a sub-appellation to put Napa in larger print on the label.

The territorialization set in process is epitomized by the marketing activities of the associations. The NVV, although a master of marketing before the AVAs, embraced the appellation and *terroir* concepts. It had used Amerine and Winkler's heat summation maps of California and the longstanding beliefs that the central valley acted as a heat pump to drawing cold oceanic air up the valley, but to further emphasize Napa's cohesiveness and diversity the NVV collaborated with UC Davis on more focused research. New understandings of climatic, microclimatic and soil differences revealed a region with exceptional variations and fluctuations. That research has now become the foundation of the associations and winegrower's *terroir*-based differentiation strategies. To make the research and valley accessible, the NVV provides winery maps, AVA maps, vintage charts, brochures, press kits and sundry other marketing materials for tourists, distant aficionados, and critics and members of the trade who pass this PR on in their publications. The NVV website is used as a hub to offer this information and to direct people onto the appellations and individual winery websites. The association works closely with private research and consulting firms like Silicon Valley Bank, Gomberg-Fedrickson & Associates, and MKF. MKF is located in Napa and a mutually beneficial relationship has been develop with MKF developing products for and with the NVV, which have been sold to other territories and wineries. Cooperation between the NVV and all the consultants reduces the costs of their market and operational analysis when sold to winegrowers. The usefulness of the appellation in wine marketing is indicated by the 85 percent of the valley's wineries that are members of the NVV and pay substantial fees that are invested primarily in promotion.

The flagship of the NVV's marketing campaigns is what has become the world's largest annual charity wine auction. The four-day event attracts wealthy aficionados and celebrities, and besides raising millions for charity, earns a great deal of media coverage. Participating members must donate wine, or host a hospitality event, and/or hold an open house at their winery. If members don't participate in the charity auction, they must donate to a barrel auction held for the trade. While the incomings from the barrel auction are reinvested into promotion, its primary goal is to generate interest in the current vintage and is an attempt to create an *en primeur* market. The auctions have been incorporated into US and international marketing tours. Proceeds are donated to local charities of the host city, but Napa is the focus of local media attention. Seminars, tastings, luncheons, dinners and so forth accompany these road shows. The media, sommeliers and other targeted audiences are brought to Napa and events organized for where individual winegrowers to showcase their wines.

Collective marketing also sustain the groups that petitioned for sub-appellations. These associations are particularly strong in Carneros, Rutherford, Oakville and a few others, but have not taken root in others like Chiles Valley or Wild Horse. Reasons for lack of association include youth of the appellation, disinterest, and acrimony arising from the AVA application. Overcoming the latter problem, the Stag's Leap association has become a vehicle for winegrowers collaborating on research, marketing and sharing of information. To that end, another Davis professor lent his analysis to the explanation and legitimization of the AVA's geographical distinctiveness. The association uses typical marketing practices such as tastings, dinners, and meet the vintner events, but innovation has required effort. The creation of an appellation collection of 12 bottles, one from each of the association's wineries, required obtaining a law allowing more than one winery's bottle to be put in the same package and respecting confidentiality of each winegrower's customer list for limited sales. Another innovation was appellation tastings for hospitality staff, allowing them to understand and recommend each other's wines.

The sub-appellations are increasingly forums for direct collaboration between grapegrowers and wineries. Grapegrowers are members of the Stag's leap association and take part in events. Voting is done on a one member-one vote basis and both sides, pay dues based on bearing and non-bearing acreage. Yet, grapegrowers may not be completely in the fold. Only one place on Stag's Leap's six-member board is reserved for grapegrowers and grapegrowers only pay a token amount of the full membership dues paid by vintners. All the same, the "pretty tight group" keeps focused on territorial concerns, leaving political or social issues to the better resourced NVVA. The two organizations haven't and don't work side-by-side, although the NVV reaches out to and supports the sub-appellations more than in the past.

## Brand Culture

Appellation, collective marketing, and even the estate winery sit uneasy within a culture where the brand takes pre-eminence. Napa's history is marked by this tension, from the domination of the CWA at the turn of the 20th century, to Gallo's control in the 1950s–70s, and perhaps most strikingly by the takeover of Napa's foremost icons in the 1960s. Inglenook was taken over by the United Vintners cooperative, which subsequently transferred brand rights to the Vodka purveyor Heublein. These firms not only leveraged the Inglenook name to sell various industrial volume wine made from central valley grapes, but also neglected the estate itself. In later acquisitions, Beaulieu by Heublein and Beringer by Nestlé, the folly of degrading a star property was corrected.

The leveraging of estate name or Napa connection became more sophisticated with time. Beringer was a long-time purveyor of a volume Non-Napa wine as Napa Ridge until 2000, but has since used the Beringer estate as the umbrella brand for a line-up of premium wines, most of which have no Napa grapes in them. Another system used by corporate wineries is to buy or lease large pieces of vineyard, and buy grapes, to make relatively volume blends, while producing estate and vineyard designate wines. The corporate presence is hidden behind a selection of formerly independent wineries and estates (e.g. Foster's has Beringer, St. Clement, Stag's Leap Winery, and Etude while Kendall-Jackson has Cardinale, Lokoya, and Atalon). The family legacy of estates may be maintained in heritage propaganda, or a similar image may be achieved through the presentation of a star winemaker.

Reputation is what brings the corporations to associate their brands with Napa and its estates. A reverse process also occurs where winegrowers who have built their names in Napa have extended into larger and more diverse operations in other regions and continents. Duckhorn and Phelps limited this expansion to North Coast appellations, but the Mondavi empire extended through the North Coast, the Central Valley and into several continents. Volume brands and estates were given a name made in Napa, and several labelled with pictures of landmark wineries or use of the same label motifs. Whether this blurring of the lines between appellation and brand has a harmful or beneficial impact on the valley's reputation or whether selling Napa associated wines cuts in on the sales of real Napa wines are open questions. However, the attempt to capitalize on this expansion, by issuing stock has claimed one notable casualty. The Mondavi empire imploded when corporate performance couldn't meet the stock market's demands for growth. In the aftermath, the valley lost its leading family estate.[58]

A recognition of the power of the brand, and the search for brand equity pervades Napa, influencing large and small winegrowers. Although the foundation for brand equity is both vertical and horizontal quality, wineries cultivate an appeal based the history of the winery, the celebrity status of the owner (e.g. Coppola and Andretti), or something more playful. Examples of the latter are Duckhorn's flock of anatidae labels and Frog's Leap's motto "Time's Fun When You're Having

Flies." A type of brand particular to the industry is the proprietary label, a tool used to overcome the American association of wine quality with a varietal and the limitations of the 75 percent requirement of the varietal labelling system. Joseph Phelps used this innovation to avoid stigmatizing his Bordeaux blend with a table wine label.[59] The strategy was picked up by the Mondavi-Rothschild venture and most notably extended as designation that could be used by members of the Meritage Society, irrespective of where their grapes originate.

The pursuit of brands, however, often puts the interests of the territory at odds, not only with external expropriation, but also with its own winegrowers and those who would play in the grey areas in between. Not only are there tensions between corporate and territorial trademark rights, the existence of both a Napa county appellation and the Napa AVA increase the means to use the Napa name without using Napa grapes. Unlike its French counterpart, however, the AVA system was not established with either comprehensive trademark protection or support for enforcement. The NVV has been made aware of these imperfections and has had to go it alone, engaging in legal battles at the county, state and federal levels to gain greater collective trademark protection.

The most complex issue derives from AVA trademarks being imposed on a pre-existing corporate trademark system and demand for protection by pre-existing brand owners. At AVA system was bereft of a mechanism for dealing with this problem at its establishment, but the use of grandfathering was precipitated by a fight in Stag's Leap. Initiated in the 1880s by the founder of Stag's Leap Wine Cellars, the name became famous after Warren Winarski's took over and won the Judgement of Paris. When Carl Doumani purchased land nearby and opened Stag's Leap Winery, the two engaged in a multi-year legal battle for exclusive rights.[60] The judge ruled anybody could use a geographical name, and after each spending $100,000, the two parties decided on subtle changes to their winery names. When others decided to form a Stag's Leap AVA, both Winarski and Doumani were opposed. Litigation against the Pine Ridge winery's use of Stag's Leap failed and the AVA took precedence over private use. The most important consequence of these battles was a BATF regulation limiting the grandfathering of wineries with the right to use the name of an appellation as the name of the estate to those that predate 1986. Territorialization was fostered as well, but where Doumani eventually joined the Stag's Leap association, Winarski never joined, despite avid support of the NVV, the Silverado Trail Wineries Association, and many common causes.

Unfortunately, the issue of pre-existing brands wasn't closed. The Bronco Wine Company, next to Gallo the largest and most successful of California's "family owned" wine producers, is the bête noire of Napa because of its willingness to pursue loopholes in appellation laws. In 1994 Bronco bought the Rutherford Vineyards and Rutherford Vintners brand names. In 1995 they used those names in the bottling of wines that contained enough non-Napa wine that they had to be labelled as red table wine and continued to use the Rutherford name in later years for non-Napa wines. Bronco also bought the name of the small Charles Shaw winery, which he would turn into the mass production and inexpensive colossus known as

"Two Buck Chuck" in the early 2000s. In 1999 Bronco received permission from Napa County to build a bottling plant with a capacity of up to 18 million cases. That amount is almost double the Napa Valley's total production of 10 million cases. Shortly, thereafter in January 2000, Bronco bought the Napa Ridge Label from the Beringer's for 42 million dollars.[61] Napa Ridge was already an issue of dissension because although evaluated as "very good" by the Wine Spectator and sold for eight to 11 dollars a bottle, its grapes were non-Napa.[62] Bronco could access its 35,000 acres of central valley grapes and produce a quantity and quality undermining the relationship between name and provenance.

The NVV first started discussed closing the grandfathering loophole in 1999; considering state labelling legislation, a "seal of authenticity" for 100 percent Napa, and petitioning the TTB to close the grandfathering loophole within 10 years. In 2000 they instigated legislation to ban mentioning Napa on a label if the wine itself did not come from Napa. The law was passed, but Bronco took issue. It went to the California Court of Appeal which ruled that it pre-empted federal law. The NVV appealed to the California Supreme Court and won. Federal Supreme Court wouldn't hear Bronco's appeal and it lost again on different issues at the California Court of Appeal. Finally, in 2006, after losing another appeal to the California Supreme Court, Bronco agreed with the NVV to put the required proportions of Napa grapes in Napa labelled products. Most of the Napa Ridge production went into a similar looking Harlow Ridge brand.[63]

The grandfather issue won't die however. In 2007 the TTB reversed its stance when winegrowers in Calistoga applied for an AVA appellation that conflicted with the Calistoga Cellars established in 1997. The TTB agrees that the vintner, who uses non-Napa and non-Calistoga grapes, should not lose the investments made in his brand with the introduction of the AVA. One irony of this Calistoga conflict is that within Napa, this territory has perhaps the clearest historical and geological basis for differentiation. A second is that Calistoga Cellars claims to be oppressed by industry giants,[64] when the AVA is a tool to support smaller winegrowers. Grapegrowers, large and small vintners, their associations, and local and state politicians lined up against this rollback.

Another problematic relic is the 75 and 85 proportions given to county and AVA appellations. At their establishment, many inside and outside the valley considered these regulations too accommodating, but there has not being a concerted effort for change at a national, state or territorial level. Napa vintners appreciate the flexibility it gives them and use estate bottling and vintage labels to differentiate. Even grapegrowers suggest that there is merit in this flexibility. Jack Stuart, a former NVV president and head of Silverado Vineyards, however, claimed that there is a legitimate argument for raising the ceiling to 90 percent. The mixed demand for a more restrictive proportion is suggested by the attempt by the NVV to develop a 100 percent Napa label in 2003. The TTB allowed the logo and the NVV to monitor and enforce compliance, but although at the outset a few winegrowers, such as Dutch Henry, signed on, the initiative didn't gathered much force. This attempt at self-governance reflects smaller winegrowers searching for differentiation tools and is a reasonable indication of their influence in the NVV with its one winery-one vote rule (although large companies may have two or three wineries).

Yet, another example of Napa's *sui generis* efforts was the previously mentioned 1990 law of conjunction requiring all wine labelled with a Napa sub-appellation use the Napa designation in larger print. The NVV won the first European recognition of an American Geographical Indication and is working with American congressmen, the TTB, the Europeans and the Chinese on the protection of its trademark. Furthermore it is working with other renowned wine territories on protecting place-based trademarks. While the NVV has led trademark defence efforts, the NVGGA initiated a county ordinance requiring any new winery to use at least 75 percent Napa grapes. They did so to stop wineries from making volume wines from trucked-in grapes and labelling them bottled in Napa. Achieving this ordinance was a rancorous battle that shook the solidarity of the vintners, and is linked to the later discussed agricultural preserve. The wineries making such wines before the ordinance are still allowed to do so.

A few ex-NVV presidents claimed that even after giving up their position they still devoted a large proportion of their time to these issues. Smaller firms are disproportionately conspicuous in these efforts to protect the appellation. Napa has not been entirely alone in these endeavours, it received help from representatives and administrative officials at various levels, including the TTB. However, until recently it has not received much support from industry associations or from other territories in undertaking these expensive and time-consuming tasks. The NVV tried to work with the Wine Institute's (WI) Public Policy Committee to propose appellation rule changes to the TTB, but the two were unable to resolve a path forward. This result could be expected given that large corporations have little interest in disrupting the blending grapes and price differences among regions and that membership power in the WI is determined by the dues paid by volume. According to Purdue, Gallo with one third of California production, exercised its power to the detriment of the smaller winegrowers. Their power led to the elimination of the California Wine Commission,[65] Gallo gaining 50 percent of USDA's California wine export subsidies at one time, and the establishment of the rival Family Winemakers.[66] Gallo is not alone, 90 percent of California wine is made by 25 companies. When the NVV sponsored its labelling bill in the legislature Bronco's owner stated "Should the governor choose to sign SB 1293, we will be left with only one viable option – to work with other impacted wine industry members to seek judicial relief."[67] Those members include his family relations among the Gallos. Subsequently the WI stated its "role is not to create further tension, which is clearly what we would have done by taking a position," said its then president John De Luca.[68]

The NVV's relationship with the Wine Institute has not, however, been acrimonious. Dawnine Dyer, (NVV President 2002) regretted the cleavage between the WI and the FWC, even though as a small winery owner, her needs are closer to the positions of the FWC. The legislators only want to hear one voice and neither the FWC, Wine America, or any other organization has the clout of the WI. They don't have the membership or production numbers to carry much weight nor a legal mandate to govern their industry. Furthermore, the interests of the big producers and the WI have aligned with small winegrowers as value growth has shifted to *terroir*-based differentiation. Yet, the ambivalence remains: The WI website warns "The establishment of a viticultural area (AVA) may not always be a good thing ... (because) ... Sometimes it can strip a winery of its brand name identity."[69] And where the WI supports the NVV in the Calistoga fight, it sided with the big producers rather than the FWC when Sonoma's applied to the legislature for similar respect for its county appellation.[70]

## Competitive Quality Control and its Limitations

A frequent denunciation of European appellations is the regulation of yields, varieties, viticulture and winemaking methods, and other forms of quality control. California provides interesting perspectives on such criticisms. It offers an exceptional example of the need for quality control, of the power of an integrated organization to impose it, and on volumes greater than any appellation. Gallo has received unequivocal praise for prodigiously investing in R&D and training and effective implementation through quality control. It raised the standards of California wine enormously. On the other hand, Napa's success is used to suggest that AOC like quality control is unnecessary. But in Napa there was no rush to convert thousands of winegrowers to fine wine. Until the 1970s, a handful of wineries were responsible for its reputation and Gallo took care of quality control for the bulk of production. Like their 19th century predecessors, that handful of producers cooperated to raise quality levels. As their reputation expanded they drew more expertise, investment and competition to the valley. Quality evolved rapidly. Yet, despite this seemingly open development, Napa semi-formalized sharing and self-regulated. Moreover it looks to external bodies for support and governance. These forms of quality control exist outside of the AVA system, but they are there.

Scientific quality control, friendly to horizontal quality, arrived in 1938 in the person of André Tchelistcheff. A French trained enologist imported to run Beaulieu Vineyard, he brought expertise in the chemistry of winemaking and an appreciation of *terroir*. The genius of his impact was leadership that fit the cooperative ethos of Napa's quality winemakers and building on the relationship with UC Davis. Somewhat surreptitiously while employed at Beaulieu, Tchelistcheff's collaboration and diffusion of expertise amplified in 1945 when he founded Napa's first enological laboratory and the Napa Valley Technical Group. Young winemakers such as Stewart, Huntsinger, Martini and Mondavi who would lead the winery renaissance gathered around, shared information, and influential tours to European estates organized. The technical group continues to provide a forum of exchange, as many winegrowers, large and small, own laboratories, conduct their own research, and have a depth of practical expertise and educational background. Competition is reduced by focusing on common objectives and open discourse. The group elects presidents and officers, but is not burdened by any formal bureaucracy. A sister organization, the Napa Valley Vineyard Technical Group was formed in the 1990s in association with UC cooperative extension and the NVGGA, offering discussions and seminars. This being California, there is faith in and rapid uptake of technology, such as remote sensing and vineyard sensor technologies to map soil differences in parcels and enable spontaneous management.

The exchanges inside and outside the technical groups are enhanced by Napa's relationship with UC Davis. The relationship dates to the 19th century, and although prohibition killed off viticulture and oenology for a period, when the programmes were resurrected in the 1930s, Tchelistcheff was there to revive the relationship. Commuting to Davis weekly, he is quoted as saying "the highest title that I hope will be buried with me and put on my grave, 'a permanent student of the University of California.'"[71] Although much of the University's work could be applied to most viticultural or winemaking practices, irrespective of the intended volume and market, there were also contributions at that early stage which supported premium production.[72] With increasing focus on horizontal quality, Davis devoted more attention to investigating vineyard and wine differentiation. The strength of the Davis-Napa relation is reflected in the dominance of Davis trained winemakers and viticulturalists in the valley, the NVV's donation of the 40-acre Oakville research station, and financial support. The Martini family gave an endowed chair and Robert Mondavi US$25 million for an institute for food and wine science. In addition to its research activities and undergraduate and graduate training, Davis provides continuing education and, with US Agriculture, an extension service in Napa.

The valley is also well populated by graduates from Fresno State and other universities, and healthy rivalry. One leading viticulturalist claimed that Fresno is California's real agricultural school. Many people take programmes or courses at Sonoma that focus on business and social science aspects of the wine industry. One interviewee described these courses as being more useful, compared to the overly theoretical approach taken at Davis. Napa Valley College provides a viticulture and winery technology program to which the NVV has contributed substantial funds. The college also offers general degrees useful to the hospitality or other services related to the wine industry. Ameila Ceja, president of Ceja credited instructors at Napa Valley College for helping her and several others succeed in the winemaking business. The NVVA and the NVGGA also sponsor technical and business seminars for members.

Despite Davis' reputation and efforts of the technical groups, R&D has been limited and under-funded, largely because government, winegrowers and related businesses are not willing to invest in the collective effort. Non-proprietary R&D, that available to independent winegrowers, is distributed throughout university departments in California and the US, and the State and Federal departments of agriculture. The overall funding level is low in comparison to other industries and wine's contribution to the California economy. Perdue, for example, cites the handful of full-time researchers at Davis versus 75 full-time scientists at Montpellier and the $7 million that the Australians spend on research.[73] This lack of R&D funding inhibited California's response to phylloxera in the 1980s and 1990s and compounded University and winegrower group denial of the problem and failure to consider French experience and advice.

On the other hand academics have established the American Vineyard Foundation (AVF) to get collective funding from industry to research. The roughly US$1 million raised annually remains small, but proportional to production, small and medium winegrowers give generously.[74] Moreover, the AVF represents an interesting model for overcoming the failure of collective investment because donators can influence research areas. The response to Pierce's disease reveals collective efforts at several levels. It is a bacteria carried by a bug called the glassy-winged sharpshooter that stops vines from taking up water and has devastated many Californian grape growing areas. When the consequences became obvious in the late 1990s, US$60 million was raised for research on the disease's biology and ecology, for vaccines, pesticides, integrated pest management systems, resistant plants and so on.[75] The Federal and State governments provided most of the money and research, while State and County monitoring systems prevent transportation and nursery based spread. Industry assessed US$5 million from within, with both the NVVA and NVGGA voting to assess all members for donations for research and other means of dealing with the disease. The same organizations mobilized with the county, their workforce, residents and visitors to recognize the dangers and report sightings.

To fight common threats like Pierce's disease, common chemical and biological treatments are often made obligatory, threatening organic practices and the unity of winegrowers. Fortunately, Pierce's disease prompted remediation of riparian habitat and removal of host invasive species. Beringer vineyards worked with the Natural Resources Conservation Service, UC Berkeley, UC Cooperative Extension, and the Department of Fish and Game to devise the strategy.[76] Subsequently, the Rutherford sub-appellation spent US$80,000 to restore its section of the Napa River. Lots of other bugs, fungi, virus and bacteria require attention, moreover research is needed into viticulture and winemaking. The NVV and NVGGA, in concert with other organizations and governments increasingly deal with these issues through self-assessments. Some are voluntary, others require legal authority to impose fees and remove the potential for freeriders. The legal precedent to impose these assessments was won by rootstock growers when a majority wanted to fund research.[77]

A crucial area where Napa is yet to introduce any collective controls is the expansion of vineyards into poor quality *terroir*. Although vineyards cover only 45,275 of Napa County's 485,120 acres, most of the land is rough mountain terrain and vineyards have steadily climbed up the slopes. In 1961, vineyards only covered 10,422 acres, but 19,953 in 1973, 28,379 in 1983, and 35,846 in 1997.[78] Some of the quality disparities inherent to this expansion is filtered by the sub-appellations, but as much of these grapes goes into generic blends there is real potential for quality problems and surpluses. Bronco's Two Buck Chuck was a testament to Napa's surplus and dilution of its brand. Brokers and merchants also source wines at the Napa, sub-appellation, and vineyard level to provide distributors and retailers with their own brands or private labels, and control over the margins.[79] Beckstoffer has warned his fellow grapegrowers of the dangers, citing instances where Napa cabernet sold at discounts to Sonoma and Lake County wine and the threat to the appellation; "maybe an entrepreneur can make a buck with bad wine at a low price, but it's a cancer to Napa Valley's name."[80]

## Value Chain

Napa's value chain is distinguished by success and frustration. Remarkable mechanisms have been devised to improve relations between grapegrowers and winemakers. In production, half of all grapes pass from the hands of grapegrowers to winemakers and flexibility in these relations is responsible for much of Napa's dynamism and its capacity to generate variety. On the other hand, not only does Napa lack a cohort of merchants committed to selling its wine, winegrowers faces both regulatory and oligopolistic control of distribution channels. The valley is marked by a pronounced division between large integrated beverage companies with their own Napa estates, wineries and large volumes, and the much more numerous small wineries and estates.

### From Vintners vs Grapegrowers to Winegrowers

Many, if not most, of Napa's wineries have started up and built strong reputations on the basis of grapes they have bought from someone else. Winemakers can purchase grapes from hundreds of independent grapegrowers or estate wineries to find the variety and qualities that suit the blends or varietals they wish to make. They don't need their own land or viticulture expertise to enter the industry. This dynamism depends, however, on a relationship where both buyer and seller are trying to optimize their gains. Market cycles that encourage over-planting and grape gluts, price inflation and deflation, plus diseases, regulations and other pressures complicate the relationship. Napa's success has in no small way resulted from developing organizations and mechanisms to diffuse this tension and allow grapegrowers, estates and wineries to make investments and co-evolve. The heretofore weaker half of the territorial equation is responsible for most of these innovations.

   Until the 1970s the valley was the scene of mixed agriculture: walnuts, prunes, other crops, raising cattle. The Farm Bureau represented these interests, including those of grapegrowers, who, like most farmers, were accustomed to paying the market price for their product. For most grapegrowers, Gallo set the market price and they were happy with the pricing offered. To the extent that grapegrowers had a collective voice it was through the cooperatives. Expectations changed with the shift of the valley towards premium wines, and the need for a separate voice became apparent. The instigation of the NVGGA came from an unlikely source.

Andy Beckstoffer brought Heublein into Napa to purchase the Inglenook and Beaulieu estates in the late 1960s. He then set up a vineyard management company to supply the right varieties and qualities of grapes. In that capacity, he encouraged the improvement of grapegrower practices and better varieties, and enticed outside investment in the vineyards. He took over the management company when Heublein divested and took on a shared interest with the rest of the grapegrowers. He formed the NVGGA with six of the valley's progressive grapegrowers.[81] Together they set to raise the status of the industry's second-class citizens.

Their first success was to reduce the information asymmetry between grape producer and buyer by initiating the crush report. They enlisted Senator Claire Berryhill to ask for legislation requiring all wineries to report purchase prices and for the State to publish them in a Crush Report. Until then grapegrowers only had wire service reports of offered prices, irrespective of whether prices were accepted and in disregard for grape origin. Now there is transparency of yearly prices according to grape variety and region, and brokers use the government service. Also, to redress this asymmetry, the grapegrowers set about resolving the wineries' ignorance of their costs. The NVGGA researched and published their members costs and then used collective bargaining and the understanding of vintners like Robert Mondavi to raise prices.

To further decommodify their product, grapegrowers took to quality contracts. These raised and stabilized prices to enable investments in vineyard and viticulture. This strategy coalesced with that of vintners seeking protection from price spikes, stability of supply, and for assurances of quality. Contracts last from a year to decades, renegotiated at intervals or evergreen, and depending on the transaction partners, the contract may be verbal or written. Terms include the diversity of ways vines can be pruned; sprayed or saved from spraying; irrigated; and when and how the grapes can be harvested. Payment by the acre encourages quality more than payment by the ton. Bonuses and penalties are assessed. Meeting quality standards requires more interaction between grapegrowers and vintners, and both generally favor the more demanding yet mutually beneficial relationship. Vintners know what they want in their grapes and may have to convince a grapegrower to change their practices. Conversely, a grapegrower may demand vintners to separate the vats of different growers and to taste and discuss what's in the bottle, enabling him or her to understand the relationship between vine and wine better.[82] These relationships don't always go smoothly and investments can be made into a relationship that goes sour or ends with the opportunism of one or both sides.

The bottle pricing formula is an initiative supported by the NVVGA to link wine price with grape price. Two relatively simple formulae allow grapegrowers to capture a fair share of the value of the end product. The price of a ton of grapes may be set at either 100 times a bottle's retail price or, for more stability, 26 percent of the vintner's FOB case price. Either way, grape pricing goes up or down with the price paid for the end product. Beckstoffer introduced bottle pricing in the

1970s and demands that his grapes must go into bottles above a certain price range. The extent to which bottle pricing formula is used throughout the valley is unknown. However, its impact may be gauged by the fact that in 2001 there were 544 different pricing schemes for cabernet sauvignon alone.

Vineyard designate wines are a more profound way to capture the inherent characteristics of the land and of a grapegrower's investments into it. Grapegrowers whose vineyards consistently produce superior grapes now request, and some can demand, that their vineyard name be put on the bottle and they be compensated for the higher value. Branding the name of the vineyard enables the vineyard to outlive a wine it is put in and the contract with the vintner. Vineyard designates also assign a lot of value to the land. An indication of vineyard designate value was the fighting over trademark rights to Kalon, a vineyard named by Hamilton Crabb in 1868. Part-owner Robert Mondavi owned half of the original vineyard and sought exclusive use because of first use and brand development, but Beckstoffer, owning the other half, argued that the name predated present ownership. Typical of long-term and collective vision in Napa, the two eventually, not only agreed to share the trademark, but to set up a registry of historic vineyards.

Perhaps the most innovative way grapegrowers capturing the value in their vineyards is by operating custom crush facilities. These multiplex wineries enable several grape buyers to become winemakers. The grapegrower provides economies of scale and scope in the various destemming, fermenting, storage, laboratory, packaging, showroom and office equipment and even personnel, while winemakers buy their grapes in the aspiration of making a high-end wine. Andy Hoxsey invented the custom crush, braving not only the substantial resources demanded, but also seeing beyond the complexities of the TTB's regulations. An early proponent of organic viticulture, he used the facilities to sustain the land and his family's heritage. Other grapegrowers have followed, cultivating new winemakers and producing some of Napa's finest wines.

On paper there are hundreds of grapegrowers in the valley, but most are very small and although the escalation of prices enabled a small minority to live off viticulture, most don't work their land. Many of those that do, have expanded their efforts to take care of other vineyards. The majority of the grapegrower vineyards, and many estates, are taken care of by their vineyard management companies. Mark Neal, for example, manages over 2,000 acres of vineyards for around 60 growers and 72 wineries. Besides pruning, these companies provide GIS systems to map the soil differences in vineyards; manage farm labour and associated bureaucratic red tape; soil amendments and appropriate clone selections; and other capacities necessary to produce Napa quality wine, but that requires onerous economies of scale. They have scaled their expertise to service the grapegrowers with small holdings, holiday home owners with a few acres, investment companies, and estate wineries. Vineyard management is one way that about 58 grapegrowers have found to not only survive, but managers such as Ceja, Abreu, Neal, and Piña to achieve renown. Remarkably, the county only made vineyard management legal in 2006. The firms had survived on the good will of the community and not causing disturbances.

Yet, despite how these innovations have improved the terms of trade, the shadow of commodity prices is long. Indeed, one grape grower claimed that the vast majority of wine is still blended and the use of vineyard designates by grapegrowers is few. Another claimed that some vintners are reluctant to use vineyard designates because of the recognition and power obtained by the grapegrower. The escalation of land prices, well-heeled aspiring winemakers and rich second home seekers have tempted some to sell out of a business of historically marginal and unstable returns, and of constantly escalating investments. The complexities of family inheritance issues add to incentives to sell. Those who remain grapegrowers must do it for the love of it, because, according to the vineyard manager, winemaker, and grapegrower Mark Neal, there is no way farming returns can equal the monetary benefits of selling. Consequently, many believe the days of the grapegrower making a living solely from grapegrowing are numbered.

The most likely evolution for the grapegrowers is to become a winegrower. Winegrowing increases the capacity to capture the economic value of the land, labour and equipment investments in two crucial ways. Winegrowing enables producers to avoid the immediacy of the harvest and its spot prices by storing value in vats and bottles this can mitigate the severe price elasticity imposed by over-supply. Price destruction can occur even when contracts are held, because many allow for cancellation in downturns. Second, bottling wine with one's own grapes is the only way to ensure that total value from differentiation can be captured.

Most grapegrowers consider it a matter of time before all make their own wine. Several directors of the NVGGA make wine and those that do not, are into vineyard management or would like to have wineries – this, according to Volker Eisele, past-president of the NVGGA, and someone who began transferring grape production to his own winery in 1991. The biggest grapegrower, Beckstoffer, disagrees with this fatalism, trusting in his ability to define ways to secure the value of his grapes and investing in technology to improve quality. He insists that not only will the pure grapegrower survive, but that they are in a better position to capitalize on the diverse qualities produced by a vineyard. Beckstoffer claims to be able to find an appropriate buyer for the varieties and qualities of his vineyard and thus has 10 shots or more at getting his grapes into a top wine. A single winemaker, owning a vineyard, has to figure out a way to blend or market several lots of wine of different qualities.

Eisele did not discount the potential of grapegrowers and vintners to find new organizational structures to combine their assets and skills more satisfactorily. And of course entering winemaking is no easy task, but requires large investments in equipment, expertise, and management. Most grapegrowers converting to wineries, have to do so from cash flow, and cannot create the instant vinous palaces built by the wealthy looking to enter the wine nobility. For some it will only remain a partial shift, a means to use up their surplus or undervalued grapes or a way to bring more recognition to their land. Winemaking also requires a radical shift in approach to business and lifestyle. Most grapegrowers do not envy the lifestyle of the vintner. Businesses and homes become open houses, a great deal of time is spent entertaining retail customers and tradespeople. Indeed, distribution and marketing require a whole new set of skills and investments.

*Size Matters in Marketing and Distribution*

Since the purchase of Mondavi, Constellation Brands is now the largest Napa Valley producer. Mondavi is but one family of brands within a portfolio that is constantly changing.

> Throughout the year, we implemented carefully considered actions to strengthen our operations, portfolio mix, structure and global team. Constantly improving the way we conduct our business and our go-to-market approach is particularly important for us to remain a supplier of choice as our customers – distributors, wholesalers and retailers – continue to consolidate around the world in competitive environments, and for consumers who enjoy our products.[83]

In 2008, Constellation dropped the lower-margin volume brands (and originally estate wineries) Almaden and Inglenook, and purchased Fortune Brands which it dismembered in order to retain Clos du Bois, the best performing US super premium. Other investments and disinvestments were made around the globe as Constellation increased the premium table wine proportion of its holdings to 85 percent from 70 percent, inline with its strategy of capturing the benefits of this fast growing segment. Internally, four business units were reconfigured into a fine wine unit called Icon Estates; the premium and super-premium VineOne; with speciality and fighting varietal wines relegated to Centerra. Purchases of producers in the US and Canadian, Europe, Australia, New Zealand, South Africa and South America have provided them with established distribution channels among very diverse regulatory and taste markets. Those are its key markets, but its products are sold in 150 countries. Along with being the world's largest wine company, it is also the third largest alcoholic beverage company and enjoys parallel economies of scope in distribution and marketing. In most markets it relies on wholesalers and retailers for distribution but owns the Wine Rack in Canada and through joint venture has access to 8,200 pubs in Britain. In the US it works with distributors on supply chain efficiencies and in the important British market it has two bottling-distribution centres. The company still purchases the majority of its wine, but employs over 9,000. Net sales had reached US$5,216.4 million by 2007, of which 768.8 was selling, general and administrative expenses, with an advertising outlay of US$182.7 million. It was also able to fund an extensive benchmark study on wine consumer preferences.

Although Constellation is outstandingly large, most of Napa's wines enjoy this type of distribution, as they are produced by wineries within groups like Fosters, Gallo, Bronco, Diageo, Kendall-Jackson, Rubicon, Pernod-Ricard, Trinchero

Family Estates and so on. The majority of these wineries own extensive land holdings and buy grapes, must, or wine, to make brands that are sold using a similar playbook. The greatest volumes are sold in Napa Valley varietals such as cabernet, chardonnay and so on, with prices ranging from US$15 to US$30. Prices escalate exponentially as reserves (select grapes), sub-appellation, and vineyard designates distinctions are added. A few companies, like Kendall-Jackson, produce smaller quantities using estate or purchased grapes, but volume sales are not necessary because scope economies of the multi-regional or global production and distribution system are leveraged for these sales. Indeed they are necessary to their portfolios.

The distribution and marketing strategies of these large firms provide leverage with increasingly consolidating wholesaling, retailing, restaurant, and wholesale sub-sectors. Retailing is split into 287,286 on-premise (restaurants, bars, etc.) and 143,864 off-premise (stores) locations.[84] Sufficient data is lacking for on-premise sales but chains like Darden with 1,427 Outlets (e.g. Red Lobster, Olive Garden), Applebee's 1,892 outlets PF Chang's 261 outlets and a great number of hotel groups suggest that purchasing is highly consolidated. In a decade large-scale supermarkets and other alternate distributors grew from 10 to 20 percent of off-premise sales, with only 5 percent of locations. Smaller chains coordinate the rest of the 140,000 off-premise locations. Gaining access to these retailers is the primarily the job of a wholesaler, but in the last 30 years wholesaler numbers have decreased by over 75 percent and in each state only two or three wholesalers determine what wines are available. Several wholesalers span several states and the 10 largest wholesalers command 58 percent of the US market. [85] Firms selling less than 500,000, according to the late Tom Shelton (former President of Phelps and of the NVV), can't influence these wholesalers. Indeed lack of distribution scale was a reason for Mondavi's takeover. Thus the integration of large companies in Napa can be seen to be part of a co-evolution, but not one resulting entirely from natural selection.

To encourage States to ratify the end of prohibition, the 21st Amendment permitted each state to designate wholesalers and use other regulations to control alcohol sales. Although this contravened the constitution's protection of interstate commerce, it was not an issue for decades. With the boom in independent wineries from a handful to over 5,000, the system has prevented producers from selling directly to retailers and consumers out of state and forcing them to use licensed wholesalers in those states. The essential problem is that in most States, winegrowers can only get one wholesaler, appointed for the life of the winery, irrespective of the efficiency of representation. Income from distributing wine becomes a guaranteed rent and supports a trend to consolidation in the alcoholic beverage distribution system and with similar ramifications upstream and downstream.

The producer, wholesaler, retailer system which typically offers many distribution efficiencies is pejoratively called the three-tier system and reviled by most winegrowers in Napa. Unable to get adequate representation, winegrowers have been fighting this system for years. The NVVA spun off the Coalition for Free Trade in 1997 to take up the legal challenge to trade barriers state-by-state. Each suit requests circuit courts to declare unconstitutional the applicable state statutes that prohibit direct shipments to consumers by out-of-state businesses.[86] The organization is supported by several wine organizations and looks to their legal advisors, law professors and others for advice. Another NVVA spinoff, Free the Grapes takes the fight to the court of public opinion. Originally the Wine Institute was ambivalent to this fight, its large corporation constituents not having much to win. As the value of premium wines and appellation wines has increased the interests of large and small companies have coalesced and the WI now devotes considerable resources to removing barriers and helping winegrowers navigate the regulatory and tax burdens of direct shipping.

Wine America and the Family Winemakers of California support the battle, and against considerable vested interest. The Wine & Spirit Wholesalers of America defends their control, claiming inter-state direct sales would provide unregulated and untaxed alcohol to minors and other deviants. Arguing that few minors buy premium wine online, free trade advocates have opened the majority of states to winery-to-consumer shipping. However, many states add registration fees and paperwork burdens while only 12 states allow retailers to ship direct. Although the WI and others provide websites to reduce these transaction costs, the increased potential for direct long distance sales has not dramatically changed the fortunes of Napa small winegrowers. They still face the tasks of differentiating themselves and getting the customer to try their product.

Some mid-level brands, those selling 50–100,000 have established themselves among distributors and chains in many states, although occasionally refocusing after more ambitious forays.[87] The great majority of Napa's wineries or estates, however, produce less than 20–25,000 cases and around 60–70 percent sell below 5,000 cases. One reason for this size limitation is production drawn from a few to a couple of dozen hectares. Firms of less than 42,000 cases also get a tax credit of 90 cents per gallon. More importantly, at this scale of production a winegrower doesn't have to find a large market. Above the 25,000 case level, wholesaling becomes more important, salespeople hired or the winemaker spends more time on aeroplanes.

Small winegrowers focused on direct sales, not only because they can't get wholesalers to work for them, but also because cutting out the middlemen improves the profit margin by multiples. For the independent winegrower with disproportionate investments in land, supplies, equipment and labour, the thick margins from direct sales are necessary to survive. There are two interrelated categories of direct sales: on-site to visiting customers and distance sales based on wine club membership and internet sales. They are interrelated because the former often leads to the latter, people visit the winery and become members of the wine club or use the internet for follow-up sales.

Tasting rooms, although charging for the pleasure, entice wine sales, along with clothing, wine accessories, maps, and other sources of income and branding.

Cooking classes, food service, winery tours and barrel tastings, and so on improve the offer. The tasting room opens the door to architectural statements designed to build an aura around the wine; as are other vinous, frivolous or fantastic attractions. For the small wine producer, perhaps the greatest investments are the time and attitude necessary to evoke the image of a family-based operation, a long history, rebelliousness, severely uncompromising standards and so on. Hospitality staff are pervasive, but expensive, and the owner and family play a substantial role in frontline meet-and-greet. Tasting rooms perpetuate wine clubs sales. When the customer goes home they can purchase bottles or cases of wine at intervals, enticed by 10–20 percent discounts. Newsletters, special events, private tours, tastings, and secret handshakes provide authentic connections to the exotic realm of winemaking. Wine clubs also determine production strategy because a portfolio of varieties to maintain member interest. The dream of most winegrowers is to have all their wine allocated, sold to select customers before its bottled. That, along with a stratospheric price, defines the status of a cult wine. With the cults, clubs ensure security of supply for the customer. Winegrowers also seek a place on the wine lists of good restaurants and discount their wines for the greater exposure and for their ego.

While the margins appear to be fat, direct sales are not cheap. Promotional staff, club membership maintenance, image investments, and lifestyle intrusion impose quantifiable and less-quantifiable costs. The Internet has made the wine club less burdensome, but address and credit card changes, family break-ups, the complexities of interstate shipping, third-party and other legal issues mean that the transaction costs of direct sales don't digitally disappear. And shipping still has to be paid. Ironically, large corporations are attracted to direct sales, not only because of the margins, but also because their economies of scale and scope reduce transaction costs. They make limited investments in authenticity: supplanting the family with a star winemaker, invoking previous family ownership, or vineyard designates. Such costs are countered by a direct sales platform for Napa and other California blends. Ultimately, the direct sales, however, depend on the valley and its organization.

In addition to its winegrowing prowess Napa benefits from proximity to the San Francisco Bay Area's millions of wealthy and sophisticated citizens and wine country is integrated into its tourism itinerary. Indeed Napa ranks second to Disneyland among California's destinations. Napa worked to have San Francisco recognized as gateway to a premier wine region and as one of the world's Great Wine Capitals. Napa benefits more from this proximity than its more spread out North Coast neighbours because of its simple road and winery organization. Most visitors drive up Highway 29 stopping in emporiums such as Coppola-Niebaum and Robert Mondavi, and one or two of the less-famous south of St. Helena. Significant numbers make it north, past Beringer's to Calistoga or east to the Silverado Trail, chasing a wine tasted in a restaurant, heard about, or looking for a discovery on the sideroads and mountain valleys. This layout brings the traffic to the tasting rooms and results in 20–30 percent of sales, even among larger wineries.

Some such as Sattui sell the majority of their wine on premises. The county and the towns in the valley strongly support this flow, as do dozens of different tourist businesses dependent on it.

The NVV provides cartography and publications that make the valley legible to tourists or for a virtual tour. Visitors are apprised of the winery providing the right mix of wine, entertainment, children and dog friendly atmosphere. Perhaps the most ingenious programs educate wine educators, writers, and sommeliers. Typically experts spend a few days in the bosom of Napa's finest facilities, such as the Culinary Institute of America, and learn from respected peers why Napa represents the pinnacle of US winegrowing. Yet, tourism is a double-sword, and an equally important role of the NVV, the NVGGA and other organizations is to maintain not only the valley's agricultural and environmental integrity, but also its quality of life.

## Environmental and Social Sustainability

If the typical dynamics of the US land market had taken their course, much of Napa county would be subdivisions for Oakland and San Francisco. The returns to development, especially after taking net present value into account, have little sympathy for the meager returns of fine wine production. Silicon Valley's obliteration of winegrowing in Santa Clara Valley and the subdivision of central Sonoma County testify to that reality. Napa's evasion of a similar fate depended not only on organizational capacities per se, but ad hoc coalitions that enabled people to collaborate on a cause and to bridge differences between their primary associations or groups, and to transform them. The environmental and social and ultimately, the economic sustainability of the valley prospered on this dynamic, but still faces many challenges. The most difficult arise from a constant and complex re-articulation of public and private needs in regard to development and private property and for social welfare.

### The Land

Napa's greatest victory is the saving of the valley. In the 1960s a conservation movement was initiated by a plan to build a four-lane highway through the Valley. A coalition of residents, environmentalists, grapegrowers, and winery people organized the Upper Napa Valley Associates (UNVA) and blocked construction.[88] This same group aligned with the County administration to make the Valley the first agricultural preserve recognized under the California Land Conservation Act.[89] The campaign for the preserve pitted the grapegrowers, led by Andrew Pellisa and a majority of vintners led by Jack Davies and the L.P. Martini against a group called the Napa Valley United Farmers, led by John Daniels and Louis Strella. The latter driven primarily by a defense of property rights. Daniel and Stralla failed in a suit that went to the Supreme Court. The preserve lowers taxes for 26,000 acres of unincorporated land on the valley floor and increases the minimum lot size to an agriculture compelling 20 acres. The benefits of the preserve were underscored in 1979 when the minimum lot size was extended to 40 acres. Later as vineyards and homes increasingly competed on the Valley slopes, they were included into the preserve and the lot size increased to 160 acres, and took the preserve size to 38,000 acres. Furthermore, to support the intention of the lot size limitation, in 1990, a law removed the right to convert agricultural land to urban uses without voter approval. The previously discussed winery definition law was also passed to stop industrialization and commercialization, and much like the preserve law generated dissension among the vintners.

Napa's self-governance exists on a tenuous basis, however. The Association of Bay Area Governments (ABAG) has a mandate to feed San Francisco suburbs: "The State Department of Housing and Community Development (HCD) does not recognize local growth limits, urban growth boundaries, or other limits on housing production as valid."[90] In the period of 1999–2006 Napa County was to produce 7,063 housing units and 1,969 were to be in unincorporated areas.[91] Furthermore, according to the State's 1974 Subdivision Map Act, a property owner can subdivide land in spite of pre-existing local preservation laws if they prove that their land had been subdivided in the past. Since, "3,414 properties have been divided in Napa County, in some cases creating dozens of parcels where there was just one before."[92] Grapegrowers and others led by Volker Eisele attempted legislation sparing Napa from this problem, but although passed was rendered ineffective and dependent on enacting a county ordinance. The latter was not achieved as the supervisors feared encroaching on property rights.

On the other hand, agriculture is not everyone's vision of conservation. With some irony, residents decry some wine industry practices as environmentally unsound, as do environmentalists from outside the Valley and growing numbers of organic and sustainable winegrowers. Pesticide use put the Napa River on the impaired water body list,[93] vineyards expanded into forests, while "ripping" mountain slopes to make vineyards caused erosion and damaged water supply and river habitats. The environmentalists convinced the County to enact a law to preserve the hillsides.[94] The preservation of riparian habit has been particularly divisive because environmentalists and County officials pushed for compulsory setbacks of 150 ft. from the river. Even environmentally friendly winegrowers have difficulty with losing large strips of their vineyards. Decreasing water supply, climate change and other issues loom.

Government intervention in environmental issues date the 1930s when the Federal government urged the States to establish local soil conservation organizations. The Napa County Resource Conservation District (NCRCD) was set up and now works with the Federal Natural Resources Conservation Service (NRCS) to offer advice on soils, watershed, and resource conservation methods. Other institutions supplying resources, information and regulations are the County's Agricultural Commissioner and its Conservation, Development and Planning Department, California's Department of Fish and Game, the universities and their extension services. It cannot be said that the winegrower organizations were particularly proactive in pursuing this support or ecological matters generally. For example they did not impose informal or formal control over what many winegrowers considered irresponsible clearing of the hillsides. That said, several winegrowers committed to land stewardship founded their own organization.

In 1995, the Napa Sustainable Winegrowing Group (NSWG) organized to promote integrated pest management, but went on to advocate for biodiversity, carbon and nutrient cycling, plant-soil interactions to reduce chemical inputs, support the definition of *terroir*, and improve profitability. The group won several grants from the USDA to support its research and educational activities. Members of the group control 23,000 acres of farmed vineyard land and over 45,000 acres of un-farmed/wild land. About 20 winegrowers are certified as organic,[95] bio-dynamics has a number of adherents, and many others are practicing less stringent sustainability. A few have installed solar panels and are committed to carbon neutrality. Perhaps the most powerful statement of commitment by winegrowers to environmental preservation is the 21 that have put over 5,000 acres of their land into the Napa Land Trust.

The activities of the NSWG and other like-minded people have been harnessed and supported by the NVVA and the Wine Institute. The NVV launched its Napa Green Land certification as a means to foster vineyard conservation while solving the impass between environmental demands for riparian setbacks and the winegrower demands for vineyard. The Wine Institute recruited several Napa Valley winegrowers and leaders from other areas to create a comprehensive guide to sustainable winegrowing. The WI's sustainable winegrowing programme and the California Sustainable Winegrowing Alliance is perhaps the best example of not only vintners and grapegrowers collaborating, but also large and small producers on both sides. Environmentally, Napa still has a long road ahead, largely because of its dependence on cars and trucks. That dependence is also a symptom of social sustainability problems.

*The People*

More than most wine regions Napa is dependent on a large labour force. Not only is the work in the vineyards labour intensive, but because of direct wine sales and of other activities, many people are employed in the hospitality services. And more than most wine regions – outside of the US – the healthcare, pensions, housing and other welfare needs of the labour force is a matter of private concern. Another particular concern of this region, because of the predominately Mexican migrant makeup, is social equity. Although in comparison, Napa's success enables it to pay its workers better than other regions,[96] it is also a lightning rod for criticism of worker treatment.

The average annual salary of a (non-owner) winemaker is $101,467[97] and they receive a package of other benefits and perquisites. Below this pinnacle any full-time salaried employees receives full medical, 100 percent-funded pension, liberal family and sick leave policies, and stock options.[98] Salaried workers number several hundred at Robert Mondavi or Sutter Homes (which headquarter operations external to Napa) and usually a few at small family-run winegrowers. Besides size; land and winery ownership, use of vineyard management, consultants and other service suppliers, competition for skilled workers and high costs of living determine who gets put on salary. In medium to large firms, economies of scale allow salaried employees in vineyard, winery, marketing and administration. In small operations, wife, husband and other relations are the core workforce, possibly supplemented by salaried expertise. Many smaller winegrowers use the health, insurance, and workman's compensation packages that are offered by the farm bureau, the Family Winemakers, the California Association of Wine Grape Growers and so on. Some grapegrowers and vineyard management companies offer benefits packages, which are especially important for compliance with regulations. Not all firms are able or concerned enough to do so however. Even larger grapegrowers and wineries are finding provision of such benefits difficult due to escalating health and compensation costs.[99]

The majority of the workforce, however, is part-time or seasonal workers who receive significantly less in salary and benefits. Hispanic workers of the vineyards, and to a much lesser degree of the wineries, are the most conspicuous. These are the people who everyone unequivocally believes the industry can't do without. Nobody else will do the work, and more importantly they don't possess the tacit understanding of agriculture. Around 4,000 people work in Napa's vineyards throughout the year, and the majority of the 2,500 Latino workers are in fact long-term employees. Many are members of the core workforces, some in charge of vineyard activities. More are long-term seasonal workers, employed from winter pruning season until harvest. Another 800 or so work short-term during the harvest.[100] Retention of fulltime and seasonal workers is increasingly important because they understand the evolution of the vineyard, even each vine, and know how to select and handle the grapes. While still a minority, such workers are offered salaried benefits, most likely in a paternalistic firm such as (pre-takeover) Robert Mondavi.

Less conspicuous are the 7,000 hospitality and related staff that work in the tasting rooms, sales offices, restaurants, art galleries and so on.[101] Although there is less information on them, they share seasonality, part-time employment and relatively lower wages with their outdoor counterparts, most likely earning and working less. The total winery tourist payroll is $90,312,000 and the average yearly earnings is $13,536. A diverse lot, most come from somewhere else and some other occupation. Retirees are an unknown but large percentage. Common characteristics are a love of living in Napa, outgoing personalities, and little original knowledge of wine. They are hired by the wineries, not for their knowledge of wine, but for their personalities and people skills and wine appreciation is taught to them.[102] They work, and to a certain extent live, in Napa Valley for lifestyle reasons. Retirees bring the added benefit of already having health and pension coverage. Younger hospitality staff work a number of jobs and otherwise figure things out for themselves.

For both vineyard and winery workers, Napa lacks affordable housing and reasonable public transport. Most workers therefore live outside the valley and commute by car. A study has confirmed that 2,500 year-round farm workers commute into the Valley from cheaper accommodation outside.[103] The commuting patterns of the lower paid hospitality staff can be surmised. Travel and housing costs combined with low compensation and benefits, result in staff retention problems. To a certain extent these problems are reduced with strategies to develop a "family culture," parties and events, wine, transparency, bilingual managers and translators, and bonuses for staying the length of a season.[104] These issues are, however, collective problems as well. Napa has to avoid being cast as a luxury industry caring little for its community.

The NVV has spearheaded the response by using its auction to support local charities. Inspired by a request from the St. Helena Hospital to chair a fundraising drive[105] and by a visit to Burgundy's Hospice du Beaune auction, Robert Mondavi saw the opportunity to benefit the hospital, improve relations between vintners and community, and raise the profile of the appellation. From 1981 to 2007 they raised US$77 million, donated primarily to health care, but also to housing and youth development programs. Due to the efforts of 1,000 volunteers, little is spent on services and nearly all money is donated to charity. Although a wonderful mix of charity and marketing, the NVV has in essence created an integrated welfare function for its winegrowers and their workforce, compensating for the lack of government provision. Most services are delivered diffusely, through community-based institutions such as the health centre, but as much of the focus is on the uninsured, it reaches a substantial number of the winegrowing workforce. Other donations such as a teaching winery and worker housing directly support industry.

Money can not simply be thrown at issues, however. In the past, winegrowers put up their harvest workers, but due to winery expansion, tourist accommodation and a disinclination to house these guests on the premises or in the neighbourhood most accommodation disappeared. To better understand the problem an NVV and County housing authority investigation found not only a shortfall of 400 beds, but also that only 16 percent of winegrowers were willing to do something. The NVV lobbied the State government for a permanent tax of US$7.76 an acre on vineyard owners to build worker housing. The legislation was made possible when a referendum among vineyard owners approved the tax and the County allowed development of lots smaller than the agricultural preserve's 40 acre minimum. Building a 60-bed dormitory was only realized, however, after Joseph Phelps overcame opposition from neighbours and donated six acres. The NVV's US$645 million had to be supported by US$1.2 million in County funds and $1.5 million in State funds.

Community support for winegrower initiatives and to preclude regulation of winegrowing activities is necessary as well and the NVV has set up programmes to increase communication and service to the community. Yet as members of the community, this support is not simply functional. The vineyard manager, Oscar Renteria, for example, claimed that many winegrowers can't live with the differences between their children's opportunities and those of the farm worker's – because of the huge income produced by Napa "its damn right that the winegrowers should do something." Perhaps the strongest indicator of social sustainability and cohesion is social mobility.

From Napa's earliest days and beyond the renaissance in the 1960s and 1970s, Napa has enabled generations of cellar rats to become winemakers. Although more difficult in today's high investment-high tech industry, such a path can lead to establishing a winery or brand with bought grapes or, as in the case of Dawnine and Bill Dyer, building an estate. Vineyard workers also take this opportunity, not only going into vineyard management. Some such as Ceja take their understanding into the winery, while others receive mentoring at wineries such as TVine. This employment mobility translates into social change as previously migrant workers now stay, taking citizenship and sending their children to local schools. At its core, what is driving Napa's mobility is the need to differentiate and the desire by winemakers to be in charge of their own operations. In large firms winemakers play a more specialized role, but in small firms they can be more hands-on and creative.[106]

# Chapter 6
# Chianti Classico:
# Globalizing Sangiovese

After years of experimentation, Bettino Ricasoli, the "Iron Baron," settled upon the appropriate proportions of three grapes to improve the wines on his Chianti estate. That was 1867, his formula rapidly came to represent not only the Chianti territory, but also a style of wine. Indeed within a couple of decades, driven by mass production and the use of external grapes, chianti as a style displaced Chianti as an origin of wine. The winegrowers of the original Chianti territory, Chianti Classico, have struggled with this dilemma ever since. In their struggle they have been hamstrung, not only by a lack of self-governance, but an imposition of borders and regulations that denied them the ability to evolve with markets. The rise of variety, of estates and *terroir*, in Chianti Classico was made complex by a particular background.

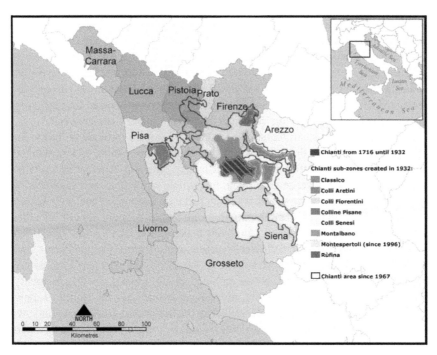

**Figure 6.1    Map of Chianti Classico among its Chianti Neigbours**

This includes a strong influence on grape choice and viticulture by industrialized winemaking and a labour force and land tenure based on sharecropping. Especially in latter decades, the development of the territory was coloured by a strain of irony, although Italy is the hearth of a vast number of the world's grape varieties, their existence, and even support for sangiovese was threatened by the popularity of the "international varieties." As the territory struggles to gain control over its quality and improve its reputation, it also has to wrestle with its definition of typicity and what that means not only for winegrowers but also for the viability of distinct varieties. An outcome of this story is a winegrower organization, that perhaps more so than any other defines the territory, indeed it has set a benchmark for territorialization.

**Territory vs Region**

Like few other territories Chianti Classico fulfils the romantic vision of a wine landscape. The territory stretches 45 kilometres north to south through mountainous terrain, from the flood plains in the outskirts of Firenze to the suburbs of Siena. Broad river plains frame the territory on the east and west. Alternating between sharp and rounded peaks and ridges, in some places the mountains reach 800 metres. The majority of this 70,000 hectares is covered by native oak and chestnut forest, but the middle elevations have been transformed by cypress and olive trees, and renovated farmhouses. As if to further demarcate the territory, *autostrades* run the length of the eastern and western borders. One enters the territory by climbing a maze of hairpin roads. The only major interior road is the via Chiantigiani that runs from Firenze to Siena. But leaving farmyards and vineyards on the valley floor, the landscape begins to hint at conscious organization. Distributed among the slopes are 7,000 hectares of specialized and increasingly densely planted and meticulously cared for vineyards. These are the vines that have the capacity to produce Chianti Classico wines.

Although the physical features of Chianti Classico provide a basis for geographical definition, the existence of the territory owes more to a prolonged collective effort. That effort is suggested by the stitching together of parts of the two provinces of Firenze and Siena, five whole communes and segments of another four.[1] That stitching together is matched by a denominational segregation of other claimants to the Chianti reputation. The fingers of more modest hills that extend to the northeast, southeast, and west make up the DOCGs of Colli (hills) Fiorentini, Colli Aretini, and Colli Senesi respectively. Rufina, Colline Pisane, and Montespertoli are more distant northern outliers from the core territory, as are a pair of areas farther south that are included in Colli Sensi. On the periphery of these areas, and in between them, are vineyards that have the right to supply DOCG Chianti. The entire area is described by the Italian and Tuscan governments and in most wine literature as the Chianti region. Few within the Chianti Classico zone would be that liberal in definition.

The winegrowers of Chianti Classico have an ample historical claim to their territorial origins. The name Chianti is ancient, possibly originating in the Etruscan family name "*Clante*" or the latin verb "*clango*" (to resound) or the name "*Clanti*" given to a piece of land donated to the monks of San Bartolomeo in 790.[2] The area associated with this name was battered together through several hundred years of political struggle. The Lombards built fortifications on Roman ruins after expelling the Empire's Byzantine successors. Afterward the area was contested by Arezzo and Siena, and then Siena and Firenze. These tensions bequeathed several castles to the present day tourism industry. The Siena-Firenze contest provided Chianti with its Black Rooster (*gallo nero*) trademark and most useful marketing myth. The *podestà* (chief magistrate) of Poggibonsi determined to settle the border in a contest of riders departing their cities at the first cock crow. The Sienese chose a well-fed and complacent white rooster, while Firenze starved a black fowl, which rose early to demand his feed. The Firenze rider was within site of Siena's suburbs before he met his late-rising competitor.

The first formal Chianti territory was born with Firenze's rise to power and prominence in the Middle Ages and the Renaissance as it extended its influence throughout Tuscany. It divided the region into leagues of semi-self-governing smaller towns. In 1250, the league of Chianti – Radda, Castellini and Gaiole was formed for defence. When the league created statues for governance in 1384, they selected the Black Rooster as symbol and with the prohibition of grape harvest before September 29th,[3] initiated territorial controls over wine production. The league governed Chianti and its wines throughout the vicissitudes of rivalry between Pope and Holy Roman Empire and of foreign incursions. Chianti's borders were first extended in 1716 when Grand Duke Cosimo included the region north to Greve and the periphery of Siena. His was one of the first legal protections of a place name. Yet in 1774 when Chianti and the other leagues were dissolved, so was this protection.

The production and reputation of the wine didn't dissolve with the formal governance of the territory. From the Middle Ages, local aristocrats or bourgeosie in Firenze, Siena or other centres owned the land, while peasants, and then later sharecroppers, worked it. Blender-merchants in Firenze collected, blended and distributed these wines, some of which found their way to the Lowlands and Britain during the Renaissance and Enlightenment. The blender-merchants were organized into one of Firenze's minor guilds,[4] a heritage shared by the present day houses of Antinori and Frescobaldi.[5] The first mention of Chianti, as a white wine, comes from the books of these merchants.

Agricultural production expanded with the industrial era and growth of urban markets. Hillsides were terraced, striated with the horizontal planting of vines, and roads were built to take produce to markets. The sharecropping – merchant-blender chain remained the dominant mode of production, with vinification ramping to an industrial scale, supplied by grapes and wine from sharecroppers and large and small landowners. In the second half of the 19th century export demand was assisted as diseases swept Bordeaux. The quantity of grapes produced in the traditional Chianti didn't satisfy this increased scale and their higher prices stimulated a search for alternatives. The extension of the Chianti name to grapes from an expanded region and other areas began. At first external bottlers printed "*chianti uso*" on their labels, but this practice soon faded.[6]

Ironically, the expansion of Chianti derives from a reputation based on *terroir*. The pre-eminence of *terroir* is suggested by the grapes, colours and styles changing over the centuries. Florentine transactions of the 14th century describe Chianti as white wine. Later, the use of the *governo*[7] process increased the red composition, but by the late 18th century red predominated, using canaiolo grapes, and lesser amounts of san gioveto (a sangiovese), mammolo, and marzimano.[8] A few large landowners attempted to make fine wine from these grapes,[9] then in the 19th century Baron Bettino Ricasoli toured France for lessons.[10] He improved viticultural methods and the equipment, practices and especially sanitation of the winery, but is best remembered for his Chianti recipe that gave predominance to sangiovese. Canaiolo was reduced to a 15 percent portion. To temper harshness, the white grapes were used; trebbiano for quick maturing wine and malvasia for aging wines.

Ricasoli's recipe succeeded internationally, but when widely adopted and liberally applied, expanded Chianti as an enologically, rather than a *terroir* defined reputation. The producer Melini's attempt to improve the traditional bottle, the presciently named *fiasco*, also went array. The tempered bottle preserved the wine better over long distance and promise authenticity, but was used irrespective of quality or origin. Indeed, it gave industrial producers a more identifiable symbol for Chianti than territory.

At the dawn of the 20th century winegrowers in Chianti faced a familiar struggle to protect their territorial trademark against fraud.

> (with exception of a few big estates like Brolio or Antinori) the grower sold his produce soon after the vendemmia to a middleman for so much a barrel of fifty litres. It was collected on the spot in uncorked flagons sealed simply with half an inch of oil, which were loaded on a special cart, from six to eight hundred at a time. The buyer paid cash on delivery, and the seller provided lunch for the carrier and food for the horses. There were two grades of wine: top-quality Chianti which could not be marketed before the following June, that is, nine months after the vintage; and the second, more ordinary wine, dubbed 'half-Chianti', which could be sold as early as February. Apart from this restriction, what happened to the wine after it left the grower's premises was entirely the merchant's affair. Which led to a good deal of abuse.[11]

Against that backdrop, a few winegrowers realized how vertical and horizontal quality could add value to their wine. They were also aware of the movements in France to create protections for appellations and how winegrowers were becoming organized.

The struggle for self-governance began in 1903 with the founding of the "*Sindacato Enologico Cooperativo del Chianti Senese.*" They were the second group in Italy, after the "*Sindacato vinicolo piemontese*" in 1902. Other smaller organizations started-up, but discord rapidly brought an end to them and the

Sienese forerunner.[12] General dissatisfaction among Italian winegrowers with fraud and the government's stance toward appellations ensured continued attempts at organization. The *"Commissione per la tutela del Chianti"* (Commission for the defence of Chianti) embarked in 1909 and enjoyed the membership of Sidney Sonnino who was briefly Prime Minister. A lasting organization was finally established in the post-WWI promise of appellation legislation. Initiated by 33 producers, but growing to 189 within a year, the members of the *Consorzio per la Difensa del Vino Tipico del Chianti e della sua Marca di Origine* (Consortium to Protect Chianti Wine and its Trademark of Origin) expected control over their appellation and authoritative tools for upgrading and promotion.

Fortunately, the winegrowers did not wait for appellation laws to begin organizing their *consorzio*. In a difficult first year, proprietors accustomed to complete autonomy had to learn to cooperate and set up an administration. Some public figures hedged their reputations by only sending representatives to meetings. The greatest threat came from the industrial producer Brambilla, who used his position on the council of the Firenze Chamber of Commerce to be obstructive and initiated a media campaign to stop the progress of the consortium. He represented other industrialists who wanted to continue to label bulk wine from other areas with impunity. The consortium withstood this first attack, and shortly achieved one of their longest lasting successes. Using existing legislation the *Gallo Nero* was made a trademark, and provided not only funding, but a label given to producers that accepted performance standards.

The consortium's hopes were dashed when, in 1926, legislation downplayed the origin in appellations, and determined that typicity to be defined primarily by a winemaking style. Shortly thereafter wine producers, industrialists and wine traders from 27 Tuscan municipalities outside the Classico area formed an opposing Consortium to ensure official recognition as Chianti. They chose a small Bacchus (*Consorzio del Bacchino* or *del Putto*) as symbol and trademark and by opening their doors to wine producers all over Tuscany quickly became representative of the largest number of producers using the name. These numbers were a powerful force when the Italian government established a commission to review its wine laws. Although the commission visited vineyards and wineries, tested soils, tasted wines, and talked to stakeholders about differences in environment and wines, it determined that Chianti was a style and not an origin. It only deemed Montalcino and Montipulciano as distinct.

The appellation laws of 1932 created a basis for territorialization, but not one necessarily approved by winegrowers in an area. The region was defined as a broad generic chianti appellation and seven sub-zones.[13] Those of historical Chianti felt denied exclusive use of the appellation and betrayed by the extension of the name to areas of Tuscany that had been using the Chianti formula or simply the name for a few decades or less and whom they felt produced inferior wine. The historical Chianti territory, itself, was extended west to San Casiano Val di Pesa and Barberineo Val d'Elsa and north to the floodplain of Firenze. There is, however, little record of resistance to that extension, as a *consorzio* concerned with internal harmony seems to have scrubbed any conflict from its written memory. As important as the boundaries, all winegrowers using the Chianti appellation had to conform to the same grape variety and quality standards. The historical Chianti winegrowers' collective ability to improve standards, to differentiate itself from other territories, was not only not supported by government mandate, it was significantly hampered. Their only compensation was the allowance to distinguish itself as the historical area, Chianti Classico.

A further imposition was the instructions that all winegrowers form a common consortium for the sake of economies of scale. Rancor between the two chianti consortiums dismissed that possibility. That the now Chianti Classico consortium lost its attempt to monopolize the Chianti name became an enduring legacy, but there was a greater implication. When France similarly reviewed its appellation laws in the 1930s, the right to self-governance by *syndicats* formed the basis of systemic governance, including establishing the INAO as an autonomous body outside the ministry of agriculture. In Italy, for the next 30 years the consortium was obsessed and frustrated with constant appeal to a politically capricious government that constantly changed, and ignored the consortium's requests.

After losing the appellation battle, and until WWII, the consortium was wracked with internecine divisions, particularly the rift between the traders and industrialists with the winegrowers, and also battles with external producers. Moderates, including the President Gino Sarrocchi, advocated immediate action to privatize the association and trademark to make the best of the classico designation. Others such as Fornaretto Vieri wrote in the newsletter "Chianti's flag will not be lowered: it stays high on top of the hills waiving our presence in order to tell both friends and enemies that Chiantigianis are not discouraged, indeed they are ready for new burdens and their souls are intent in irresistible efforts to fight back without any biased feelings, let's be clear on this, but motivated only by their hard and fecund work that already gave our soil world fame."[14] However, the devastation of the war and the subsequent decay of the sharecropping system focused the consortium on technical assistance to restore vineyards, legal assistance for war damages claims, and procuring fertilizers and fungicides. It pushed on with marketing its *Gallo Nero* trademark, while appealing to successive governments to return exclusive use of the Chianti name.

Common cause grew among winegrower groups, in the post-war decade. Frustrated Piemonte and Veneto producers joined the pursuit for improved geographical appellations laws. In Chianti Classico, president Luigi Ricasoli-Firidolfi, descendent of the Iron Baron and proprietor of the Brolio estate, while still seeking return of Chianti's exclusivity, looked for cooperation among the producers in the other Chianti areas to obtain self governance and territorial distinctions within the new framework. The other Chianti consortium resisted and pushed to rescind all distinctions within Chianti. It only agreed to perpetuate the Classico distinction when the government threatened to go ahead without waiting for winegrower agreement. Still, the law was blocked by industrial interests who rallied support from the ministry of industry and southern producers accustomed to selling their wines into Chianti blends. It didn't help Chianti Classico's case that Dalmasso, the government's leading expert for 20 years, dismissed their claim to fine wines and supported enological use of the appellation. On the other hand the consortium looked for government support to rebuild quality vineyards on hillsides, and at a cost significantly higher than the mass consumption vineyards in the valleys.

Eventually, foreign impatience with Italian wine frauds moved the government to bring in appellation legislation in 1963.[15] More consortiums organized in the Chianti region, and eventually they were able to compromise on distinctions set out in the 1932 laws. The appellation system followed the then French model, but allowed a proliferation of *Denominazione di Origine Controllata* (DOCs) with dubious *terroir* and producing high volume low quality wines. Indeed the law encouraged industrial uses amalgamating such zones into the DOCs of areas formerly considered the origins of fine wine.[16] Although allowed exclusive use of the Chianti Classico name, the consortium felt betrayed by the legislation not only because it denied it any self-governing capacities, but because it perpetuated an extended Chianti zone with one set of enological standards. Unfortunately, these standards used the most liberal Chianti recipe. It allowed 30 percent white grape use, rather than the version minimizing white grapes for aged wines.

Denied the ability to set quality and bottling location standards necessary to a quality wine, even the existence of the consortium was threatened. Agreement of at least 20 percent of all Chianti producers was needed to set up a separate consortium to apply for *Denominazione di Origine Controllata e Guarantita* (DOCGs) appellation with its better quality controls and status. The Chianti Classico consortium's path was blocked even though it gained some support from new consortiums representing other Chianti sub-zones. It was up against the much larger Putto consortium and the leader of the national appellation committee who dismissed Chianti Classico's fine wine claims. On the ground, the competition resulted in "the battle of the road signs" as each consortium tried to stamp its signature on the territory and tourism.

Capitulating to the inevitable and with the promise of some differentiation, for not appealing the DOC decision, the consortium changed its name to "Consortium of Chianti Classico Wine" (*Consorzio del vino Chianti Classico*). It set its sights on a DOCG believing that this would enable them to create higher standards. The new standard, however, would have to be built upon the DOC shared with the rest of Chianti. For that reason they worked with other Chianti consortiums on the application, while at the same time seeking differentiation. The consortiums also collaborated to improve the quality control systems entrusted to seemingly disinterested regional authorities. In 1984, after almost two decades of trying, and despite neighbouring Montepulciano and Montelcino receiving their DOCG within a few years of applying, Chianti and Chianti Classico received their DOCG. All the Chiantis could put their sub-appellations and alcohol levels on their labels, and Classico was allowed a half percentage higher.

Denied recognition and authority within the government's system, the consortium achieved quality control and promotional goals by controlling access to its trademark. These activities are detailed later, but the most dramatic result was dividing the consortium in two. This was done because the standards of the DOCG were not as high as those sought by the consortium. The *Consorzio Chianti Classico* became a quasi-government controlled organization that ensured wine met the standards of the DOCG and worked to improve the technical standards of the winegrowers. Anyone meeting those standards could earn the chianti-classico label. The provincial Chambers of Commerce, which also register the vineyards, perform the actual testing of the wines. The newly created *Consorzio Marchio Storico Chianti Classico* (Consortium of the Historic Trademark, but usually called Consorzio Gallo Nero/Black Rooster) was a voluntary organization. It tested the wines to the higher standards of the Black Rooster label, but was primarily a marketing organization. Wines that passed inspections of both consortiums could put both designations on their bottles. This solution was achieved despite being a pro forma response to self-governance limitations and different strategies amongst members. The structural differences, however, would become vehicles for greater divisions.

### Differentiation and its Discontents

Until the late 1960s, the consortium was trying to create a fine wine territory, simply on the basis of *terroir*, and without significant estate bottling. That situation changed dramatically through the 1970s and beyond. Today, large companies still dominate the territory. Thirty-seven percent of the territory's wine goes into volume brands supplied from the bulk market and from the large company's own extensive vineyards.[17] A small proportion of the bulk wine comes from many tiny producers, who while the majority of 1,150 winegrowers (40 percent) only supply 4 percent of the bulk wine production. The majority of bulk production is surplus sales from integrated producers (30–35 percent) and the cooperatives (5–10 percent). The cooperatives directly sell 17 percent of the territory's total. Winegrowers bottling some proportion of their wine number about 360 and their direct sales account for about 46 percent of the territory's production. The estate winegrowers are a diverse lot, some the scions of ancient families or former sharecroppers, large corporations, native and new to the area, with some newcomers fuelling their winegrowing passions with income from other professions. The ability of this differentiation to flourish required further struggles within the territory and with its external regulators.

The failure of the DOCG to signal quality and the inability of the consortium to change its standards to the level of fine wines forced many winegrowers to take their best production out of appellation. Limited by the the DOCGs regulations in the quest to produce a superior wines, they used a simple table wine designation. The press, especially the American, labelled these wines the super Tuscans. Sassicaia, in neighbouring Bolgheri, initiated this trend by importing Bordeaux grapes and expertise, but it was a new estate in a new winegrowing area. In Chianti Classico, the super Tuscans not only removed locomotives (or their best production) from the appellation and its reputation, but also cast doubt on the inherent quality of its signature grapes.

Many producers chose to use only Bordeaux or other international varieties in super Tuscans, while others blended them with their dominant sangiovese. Ironically, pure sangiovese was against regulations, and Montevertine was the first producer in the Chianti Classico territory (in 1977) to gain a great reputation for its Super Tuscan. Thus, Chianti Classico was faced with the embarrassing situation that one of the highest reputations in the area using the native grape was outside of the consortium and appellation. Most producers who made a super Tuscan, however, continued to produce Chianti Classico. Tension was generated in the consortium because of the different approaches, but without control over its regulations, little could be done to resolve it. A riserva classification within the Black Rooster standards did provide winegrowers with a distinction for longer aged, superior wines.[18]

The rise of the super Tuscans makes no impact in the history of the consortium,[19] but every wine writer and winegrower attests to how the consortium and DOCG system were pressured to change their regulations.[20] Meanwhile, the market for quality wine was growing and the demand for plonk sinking. Winegrowers, old and new throughout Italy, responded with improved production and a switch from bulk to estate production. Concomitantly, they became more interested in territorial differentiation and establishing appellations. Where Chianti Classico once fought a lone battle and was often characterized as arrogant and intransigent, increasingly other territories saw common cause. Informal cooperation amongst consortiums and regional federations of consortiums began in the early 1970s and by 1979, a national federation, Federdoc, was created and became a powerful lobby for greater self-governance. The rise of collective action is seen in the support given to Chianti Classico by the other Chianti consortiums when new appellation laws where promised at the outset of the 1990s.

The homogeneity and devaluing that the 1963 appellation laws imposed on Italian wine became apparent to everyone, thus despite continued resistance from the industrialists, in 1992 the new agriculture minister Giovanni Goria succeeded in replacing it.[21] The government did try to control the consortiums via allocating power to an inter-professional bureaucracy operating through chambers of commerce, but Federdoc and the consortiums repulsed the attempt. Decisions over who could apply for DOCG and the newly created IGT, however, remained with the ministry of agriculture, with the advice of the National Commitee for the Defence of Denominations of Origin and Geographical Indications (*Comitato Nazionale per la Tutela e la Valorizzone delle Denominazioni di Origine e delle Indicazioni Geographica Tipiche dei Vini*, CNDOIGTV). Unlike the French INAO however, winegrowers committed to adding value to their piece of land are in a minority the CNDOIGTV. In addition to a few places occupied by Federdoc, the committee represents big producers through the UIV (*Union Italiana Vini*), dealers through Federvini, bulk producers of a left wing disposition through the CIA, bulk producers of a right wing disposition through COLDIRETTI, another organization of producers in CONFAGRICOLTURA, and two organizations for cooperatives, the ANCA-Lega and CONFCOOPERATIVE. Although are not necessarily opposed to appellations and estates, historically these organizations have been driven by conflicting ideologies. They add significant diversity of opinion to appellation oversight.

The outcome of the 1992 appellation law, the so-called Goria's law, was not only to create a framework for appellations. As importantly, it recognized the right of consortiums to govern their appellations. Chianti Classico grabbed the opportunity to create its own regulations swiftly and throughout the 1990s would make two more attempts to change them in accordance with the demands of the market and its winegrowers. In 1984, with the onset of the common DOCG regulations, 10 percent of international or other grape varieties were allowed into the Chianti recipe and the white grape minimum reduced to 2 percent. With autonomy in 1996, Chianti Classico removed the white grape minimum and allowed foreign grapes to rise to 15 percent. In 2001 the rule changes removed all white grapes by 2005, allowed up to 20 percent foreign grapes and pure sangiovese wines. These were not easy changes to agree upon and implement. Although increased allowance of foreign grapes appeased some producers, vineyards do not change overnight and without expense. Those who could no longer include white grapes in their blends faced significant loss. And as discussed in the QC section, these rules require enforcement.

The consortium delivered a Vin Santo DOCG that provided winegrowers with an outlet for their white grapes, and there were other IGTs and DOC they could use to lessen their losses. Perhaps the most controversial outcome of the changes in rules was the impact on Chianti Classico's typicity. Many lamented the inclusion of foreign grapes and the pure sangiovese (rather than the traditional blend) as an assault on the territory's traditions and identity and as a short-term fix. They believed the sustainability of Chianti Classico's position on the world markets requires strengthening its differentiation.

Despite the efforts of the consortium to accommodate differences among its members through rule changes, at the turn of the millennium many more of the territory's leading producers left the *Consortium Marchio Storico* and considerable tension existed in both consortiums. The tensions derived from the quality control and marketing approaches taken by the marketing consortium in particular and are discussed next. Recently, however, the tension has deflated because after close to a century of trying, the consortium obtained complete authority over its appellation. The *"erga omnes"* law was brought in 2005 and provides a consortium with the power to bring in and enforce regulations on the use of the appellation, as long as its members represent a majority of users of the appellation.

## Quality Control on Paper and on the Ground

When the Chianti Classico winegrowers assembled in Radda in 1924 to found their consortium, some winegrowers were already bottling and aging a small proportion of their production. They believed they and their territory could produce wine that could be respected for their vintage, would improve with aging and should be valued accordingly. They knew that this distinction was not based solely on their *terroir*. From the outset the winegrowers committed themselves to collective action on technical advances and vertical quality control. This position was critical to their market power for several decades, even while they remained primarily providers of bulk wine for industrial wineries. They were faced with a lot of work to achieve these goals, in the vineyards, in the wineries, and with the workforce.

For centuries sharecropping shaped the landscape. Grapes were the main cash crop combined with olives, grains, livestock and various subsistence activities. Vines were interspaced among the crops and grown in the promiscuous (untrained) style. Field maples were used as support in place of trellising and different varieties grew amongst each other.[22] The different varieties were usually harvested together, irrespective of differences in ripeness. The promiscuous technique survived the expansion and rationalization of 19th century and was deemed the most appropriate means to replant after the ravages of phyloxerra in the late 19th–early 20th centuries and even after WWII. Promiscuous cultivation enhanced hydrology, crop rotation and health of the soil. It was preferred by the conservative sharecroppers, and conservative landowners saw them as the only form of labour until post-war depopulation. Still the winegrowers and their academic advisors knew that specialized vineyards produced better wine and allowed for mechanized and less costly operations. Most, but not all, of the promiscuous vines were ripped out in post-war reconstruction as Chianti expanded vineyards and mechanized for industrial production. The renovation of cellars waited until the 1970s when Chianti Classico entered into an estate driven phase of development.

The "Iron Baron" was thus monikered for the manner in which he forced his sharecroppers to adopt more advanced techniques. The consortium could not rely on such coercion, but did control access to its trademark and marketing. Until the reunification of the territory's two consortiums in 2005, use of the Black Rooster depended on meeting consortium standards. At the outset, the tests included assessments of vineyards and cellars. Just how this was done is unclear, but the standards were self-imposed and self-enforced, and met with acceptance. In contrast winegrowers in Chianti Classico and other consortiums were unhappy with the slack inspections undertaken by the various authorities policing the DOC system. In 1972 the consortium upgraded its research centre to improve monitoring for its label and enable winegrowers to test their grapes and support winemaking and aging. In time the lab fostered some of Italy's foremost oenologists and winegrowing research. In the second half of the 1970s the consortium also introduced the so-called "*Standard d'annata*", an analytic and organoleptic procedure that allowed it to compare a wine with samples representative from a given area.

Over the decades, the consortium provided technical information via written materials and by sponsoring forums of professionals and academics. The forums disseminated information and allowed comparison with territorial colleagues and those of other territories, often with lively debates. In the post-war years it lobbied for reconstruction money and helped winegrowers apply for it. The difficult task remained to convince authorities to invest in rebuilding the classico vineyards when the costs were substantially greater than rebuilding vineyards on the plains. Government experts still didn't believe that the classico area could produce wines of high enough value to justify the investments. Eventually, the government and the EU provided hundreds of millions of lira, and without which the vineyards would not have been rebuilt.

Another collective push for quality came from the initiation of cooperatives. The first, *Agricoltori del Chianti Geografico*, began in Radda in 1961, followed by *Cantina dei Castelli del Greve Pesa* in 1965, *Le Chiantigiane* in 1975 and *Cantine del Chianti Storico* in 1978. Initially, the consortium viewed cooperative membership with some trepidation because other fine wine territories had difficulties with their inclusion. It believed that because of Chianti's special environment and local economy, the requisite investments for high quality were incompatible with cooperatives. On the other hand cooperatives allowed smaller members to overcome expenses and the organizational complexities of aging, packaging and marketing. In any case, as the DOCG could be used by anybody that met its standards, the cooperatives established themselves. In short order cooperatives proved a large step forward in quality among small producers and attained the right to use the Black Rooster label as well.

Although post-war restoration created many present day vineyards, further improvements were necessary as winegrowers aspired to estate wineries and required a territory with a commensurate reputation. Several issues needed collective management to accomplish their objectives and achieve the necessary harmony among winegrowers. First, the clones (or sub-variety) of sangiovese and other varieties planted during reconstruction were of poor quality. This unfortunate development is explained by a lack of knowledge on the part of old and new landowners and among the paid workforce that supplanted sharecropping, plus a lack of concern from the nursery that supplied the clones.[23] By the 1980s the producers and the consortium's technical team realized that the vines would have to be replaced soon, both because they were nearing the end of their productive life and because quality needed improving. That was the impetus for the Chianti Classico 2000 project.

The project spanned 16 years of research to select or develop the most appropriate clones of the Sangiovese, Canaiolo, Colorino and Malvasia Nera grapes used in Chianti Classico. The clones were selected for their enological value and for their appropriateness to local conditions and agricultural efficiency. Rootstocks, planting densities (particularly to determine limitations to increases), vine training and soil management appropriate to different sites in the territory were investigated. Eventually, seven new sangiovese and one new colorino were selected as appropriate and allowed for use. Sixteen experimental vineyards totalling 25 hectares were planted. Progress was correlated against information from 10 meteorological stations and five cellars were used to determine the results. The project was designed and led by the Chianti Classico Consortium with the sanctioning of the Ministry for Agricultural and Forestry Policies and the Tuscany region, but it was paid for by the European Community. The consortium's scientists worked with the Agrarian Sciences faculties of the University of Florence and the University of Pisa. The project was also supported by private firms supplying viticultural and vinicultural products.

The Chianti Classico 2000 project is a benchmark for collective action on R&D. It produced significant results for the territory and sub-zones within it. Furthermore, it was distinguished by winegrower participation and expertise, showcasing the advancement of the estates. Fifteen estates donated land and other resources to the replanting programs. Some, such as the respected Isole e Olena were not members of the *Consortium Marchio Storico*, but were strong supporters of the technical mission of the Chianti Classico Consortium. Even Antinori, although not a member of either consortium, shared its research results with the project. The project and the consortium were not the sole location of research or cooperation. Experiments had been conducted at San Felice for decades and results shared with others in the territory.[24]

Ambitious estimates put replanting of Chianti Classico at 30 percent complete by 2000 and completed by 2010. Replanting of vineyards, however, is expensive, for the materials and equipment, and most importantly, labour. Although several wealthy people bought estates as lifestyle accoutrements, most winegrowers live off their wine. As a result, investments in vineyards and wineries are disparate. Furthermore there is the potential for freeriding with underinvestments while the general standard of Chianti Classico rises. In particular the problems of promiscuous plantings and the varieties and qualities of vines planted in the reconstruction era have and continue to pose quality problems for the territory and divide the winegrowers. The propensity of international stars to drop out of the *Consortium Marchio Storico* was partially explained by the failure to impose tougher QC. These winegrowers complained that nobody who had passed the tests of the technical consortium ever failed the supposedly tougher controls set by the marketing consortium. Demands for tighter QC were general among most winegrowers before 2005. Occasionally, winegrowers complain about inferior and cheap Chianti Classico in German supermarkets and of wine entering the territory for blending.

The core of the problem was that new regulations (1967, 1984, 1996, and 2001) continually changed the composition of the wine, but the vineyards couldn't change at the same speed. The 1984 regulations, for example, increased the sangiovese proportion, but their vines could not miraculously appear in the vineyard. Winegrowers received some flexibility to continue to use less respected reds such as Caniolo and the eventually expelled whites (trebbiano and malvasia) could be mopped up in a new vin santo appellation. Despite these allowances grapes that should not go into Chianti Classico are still blended in. It is difficult and expensive operationally to separate them in record-keeping, and accordingly entice opportunism. The problems arise from the remaining proportion of promiscuous vines, some of which are registered and some of which are not. The un-registered vines are known as the "*superficiale virtuale*" by the province and Chamber of commerce. Even among vines registered as specialized, the varieties are not recorded. Some producers only harvest differently on paper, while the authorities were satisfied with their written declarations. The proportion of unregistered to registered vines varies greatly among the different landholdings.

Another ambiguous area is the registration of vineyards for potential production. Although the *consorzio* and most winegrowers want only the real production of appropriate grapes designated as chianti classico, winemakers were allowed to use grapes/wine imported from other areas to make up their potential production. More dramatic is the fraudulent importation of wine for blending. Importing southern wine to add colour was a long-standing practice, and despite being forbidden, winegrowers and writers allude to its continuation. Italy's largest wine scandal erupted in Chianti in the early 2000s, and suspicions of fraud in Chianti Classico were confirmed in 2005. Piero Conticelli, a Florence merchant owning 200 hectares of Classico was arrested and 70,000 h/l of Chianti Classico, or around one quarter of the territory's production, confiscated. Other Chianti Classico producers were investigated. The biggest customer Ruffino had 2.5m bottles impounded because it bought 12,000 h/l of the illicit wine.

Solving these ambiguities and their divisiveness propelled amalgamation of the two *consorzii* in 2005. In the words of the Consortium's managing director Giuseppe Liberatore "the failures of individuals had extremely grave effects on the images of the other producers."[25] Fortunately, this was the attitude that the dropouts from the Marchio Storico consortium wanted to hear as reflected in their demands of the newly merged consortium. A letter sent by Felsina, Isole e Olena, Monsanto, Poggerino, Riecini, San Giusto a Rentennano and others asked the consortium to do something about "the increasing jammification of wines by the illicit blending of wines or musts from outside the zone, or even the region, using international grapes or even non-authorized varieties like Montepulciano, Negroamaro, Primitivo or Nero d'Avola and the lack of clarity as to the nature of Chianti Classico due to the overuse of authorised ameliorative grapes' (Cabernet, Merlot, Syrah)."[26]

The consortium had been preparing to deal with the first of these demands, and shortly after the merger put in place a comprehensive quality control system. In the vineyards, investigations match vineyard registration to vineyard output; ensure viticulture is up to standard on things such vine training and densities; and growing season inspections to ensure maximum yields are reduced through appropriated viticulture. The *consorzio* surveys the age and types of vines in a field to determine what actual level of production should be allowed (rather than the potential allowed by acreage). It surveys 10 percent of vineyards each summer to determine quality, need for pruning and therefore what level of quantity should be allowed that year. Inspections in the winery include those for vinification and storage and to check the accuracy of stocks and flows with records. Bottling records, whether on estate or in merchant winery are also inspected. The comprehensiveness of the system allows complete tracability by a consumer, the consortium, producer or anybody else through in value chain. The consorzio's QC activities are particularly important because quality contracts are not used to any degree in Chianti Classico. Without the arbitration of collective QC, willingness to invest in quality would decrease, while transaction costs among winegrowers and wineries increase, and the territory's reputation would suffer.[27]

The consortium is in control of these production parameters, however, it does not control either the selection of *terroir* or the definition of typicity. The Tuscan government fought for and won the right to control the EU's allocation of planting rights under the EU's laws for subsidiarity. Following in that logic, it handed over control to the provincial governments who make up a three-year plan for distribution. Some Denominations are allowed new vineyards and some aren't – new DOCs are allowed, but the DOCGs are closed. Therefore, normally, a producer can only open up more land after buying planting rights from somebody else in the same DOCG, who must pull their vines. The provinces evaluate proposed vineyards by using aerial photos and soil charts. Winegrowers ridicule this combination of arbitrary swapping of land and a method of evaluation that gives little consideration to a vineyards *terroir* potential. Furthermore, although Chianti Classico has remained remarkably limited in size for several decades, new areas can be opened with the permission of the consortium and province. However, the consortium's actual control seems to be limited in this regard – in 2005, 600 hectares were opened even though its then President Ricasoli-Firidolfi voiced opposition to the landscape becoming marred by a monoculture of vineyards.

The consortium's control over the final testing of the wine and what is considered typical is also limited. While the consortium's lab does the chemical analysis, the Chambers of Commerce, Industry and Trades in Firenze and Siena run the tasting panels. They make up tasting commissions for 11 different regions in the province, of which six are for the Chianti appellations. All tasters are registered on a list, comprised of oenologists and experienced people who in turn are chosen by a commission in Florence comprised of oenologists, agronomists and other experienced people chosen by the consorzi and the office for the repression of fraud. A list of several dozen experts is compiled for each area, from which a commission of five members is drawn. The president and secretary are always the same but the others revolve by session. From 20 and 29 sessions are held for each of the different areas, with each session testing about 12 samples. One person from the chamber of commerce and one from the *consorzio* collect and keep samples anonymously. Producers can produce these samples vat by vat or all the vats together and it is the bottler who asks for the testing. Wines are graded as either: *Idoneo* – acceptable; *Rivedible* – revise for small defaults; or *Non idoneo* – fail and reasons must be given for the evaluations. In a typical year, Chianti Classico had ten revisable and two failures.

Wines are evaluated for looks, smell and taste, each according to five criteria, and the cause of faults specified (biological, chemical, accidental, and congenital) and commented on. Somewhere in that process, the typicity of the wine is also determined. Due to the commission being made up from external experts, however, the consortium and winegrowers do not have control over what is considered typical. Although there is no consensus what typicity means in Chianti Classico, some winegrowers have been outraged when their full sangiovese wines have not been accepted when cabernet/merlot blends have been.[28] On the other hand, the panel of external experts precludes the criticisms that were directed at the French tasting panels. Composed of winegrower and négociants, those committees were said to be self-serving, back-scratching and unresponsive to market trends. Furthermore as the Tuscan panels are organized and run by the chambers of commerce, they reduce time and expense burdens on the winegrowers and consortium.

Several winegrowers attested to the power that the consortium's quality standards and quality control have had on raising the level of quality in Chianti Classico. Another very important influence was the role of the consulting oenologist, particularly as Chianti Classico is dominated by new proprietors with little or no background in wine. Two of the most influential, Maurizio Castelli and Carlo Ferrini, started out at the Consortium (Ferrini initiated the Chianti Classico 2000 project). Others migrated from other areas of Italy. Tachis and other winemakers from Antinori have been influential before and during their consulting careers.[29] The Chianti Classico estates also hire and give credit to agronomists to higher degree than estates in other regions. The downside to the use of consultants is that they tend to take similar directions on different estates. De Marchi of Isole e Olena, while acknowledging the important contribution of the consultants claimed it is dangerous to both estate and territory for the oenologist to make the decisions that break the connection between family and *terroir*. Several others held similar concerns.

The quality role of the cooperatives has evolved to meet the demand for *terroir* and to capture its added value. *Agricoltori del Chianti Geografico* and *Castelli del Greve Pesa* both demand and subsidize the replanting of vineyards, set viticulture standards and employ agronomists to teach and monitor the activities of members. When members get too old, find other employment, or otherwise lose the capacity to tend their vineyards, the coops provide the workforce to take over in the vineyard. Cooperative members are compensated for vertical and horizontal quality differences, the latter based on selections from different villages, properties and vineyards. These differences are put on the label. The emphasis on quality is such that *Geografico* claims to no longer provide a social service, it selects its members to win in an increasingly competitive market.

Quality control was perhaps the primary incentive for merging the consortium and the improved standards and controls are strongly supported throughout the territory. The administration of the consortium and the winegrowers thought that the merger would bring the dropouts back into the fold – as they would be forced to comply with the regulations, rejoining would give them a say in their creation and implementation. Things worked out much as expected. Shortly after the merger, membership of the consortium increased to 600 members representing 95 percent of the production of Chianti Classico. In 1992 membership on in the Consortium Chianti Classico represented 82 percent of production and membership in the *Consortium Marchio Storico* represented only 50 percent of production. The rejoining of Marco Pallanti of Castello di Ama, one of the most respected estates signalled a new level of cooperation. Pallanti had been a vocal critic of the consortium, but shortly after rejoining assumed its presidency.

Before the merger, it was debated whether the *Gallo Nero* trademark would survive alongside the Chianti Classico marque. Those within the consortium believed that there was too much history, investment, and power in the Black Rooster to abandon it. The two are now used in conjunction, but unlike before there is only one set of quality controls. Still a few such as the large merchant-producers like Antinori and Folonari, and a few small estates such as Montevertine remain unconvinced as to the trademark's benefits.

## Marketing the Tuscan Sun

The Chianti Classico territory provides marvellous marketing material. The backdrop of hills and forests provides a natural setting while and the vineyards and farm houses provide a landscape tamed for tourists. Castles, churches, ancient towns and villages bring provide humanity and authenticity. The interspersed wineries, bed and breakfasts, and excellent restaurants give tourists many good reasons to acquaint with the territory. As enticing as this landscape is, it doesn't market itself, nor are the castles and churches and the land preserved and improved without collective effort. The consortium and other organizations have developed a complex and interrelated strategy for marketing territory. They have used the wine as means to draw the territory together symbolically and functionally, while providing an identity and palpable atmosphere that draw people to the territory and distinguish it in external markets. The marketing effort is important to all the winegrowers, but not to the same degree and not in every aspect. Nor do all agree to share the cost of this territorialization. These variations, in demand for territorial marketing, create the biggest source of contention within the territory, its consortium, and other organizations. Many of these divisions are structural.

Producers in the territory can be divided into grape and must producers who supply the cooperatives, the larger winery-merchants, and the estates. The latter two also have extensive vineyards they integrate with their vinification and sales operations. A couple of large integrated firms – Antinori and Folanari – eschew membership in the consortium, reportedly dropping out to avoid paying dues that were directed at collective marketing. These firms emphasize their own brands, with a full range of estate and large volume blends, have large scale and complex distribution systems to sell on national and international markets and use a variety of distribution outlets, including supermarkets. Several estates as well prefer to devote their resources to supporting their own brands. These estates have international reputations, devoted costumers and regularly feature on the pages of the Wine Spectator, other magazines, websites and wine blogs. Typically they have one or a few agents in regional markets. Most of the estates that dropped out of the consortium, did so partially because of this brand focus, but with improved QC, the merged consortium's promise of a more coherent image for the territory reduced their opposition to collective marketing. Yet, not all have been convinced.

The tension between collective and private marketing is more subdued amongst the cooperatives and some firms that are similar or greater than Antinori and Folonari in scale and operations. Until about 20 years ago the coops sold the majority of their production to merchants. Since, they have developed as umbrella brands in their own right, offering a range of regional wines and in some cases buying from other Tuscan areas. The coops sell through a diversity of outlets, selling the majority through strong internal distribution systems and also succeeding internationally. Chianti Classico is an important part of the product range of the two firms claiming to be Italy's largest wine companies. Zonin and GIV possess large Chianti Classico properties and winery capacities that augment their estate, regional and varietal products. Zonin bought Castello d'Alboa in 1979, and now owns 850 hectares in the surrounding area, of which 157 hectares are vineyards. It produces a Chianti Classico Normale, Riserva, Grappa and Vin Santo, including some vineyard designate bottlings and also makes a Super Tuscan and an IGT Chardonnay. GIV owns Mellini, Serristori and Machiavelli. Mellini is the more storied of these operations because the founding family created the fiasco bottle and in 1969 it produced the territory's first single vineyard bottling. Mellini is also headquarters for production of a wide range of regional and varietal wines including Chianti, and uses quality contracts with its producers.

Besides Antinori and Folanari, other Chianti large-scale producers, originating as buyers and blenders of wine, now supplement their products with estate wines. Like Zonin and GIV, they generally support the consortium, but get limited utility from it. There are about 30 of these "industrial and merchant houses and their wine estates," to use the consortium's words.[30] Piccini accounts for 15 percent of Chianti Classico production and, as one of the world's 30 top producer-exporters, is capable of dealing with its own distribution. Yet it has supported the consortium, the DOCG and its quality controls. Coli is not as large and although specializes in territorial blends for mass distribution, has developed estate and other speciality wines to serve market demand. It was one of the first in the consortium and supports it, while claiming limited benefit from the organization's efforts. That territorial efforts do not seem substantial in these large firms' view is, however, most likely a reflection of their ability to develop other marketing and distribution efforts. It is a relative rather than absolute indicator of marketing value. All the firms, for example, use territorial descriptions and imagery in their own marketing. Still, the primary beneficiaries of territorial marketing are the estates, especially those yet to be anointed by the Wine Spectator.

Smaller, less-renowned estates have difficulty finding distributors in Italy because the market is fragmented and big producers control the distribution channels. More wine can be sold to foreign buyers and faster, but sales to Italian restaurants and wine shops are coveted because they require Tuscan wines to be on their menus and are stable customers. Furthermore, building an overseas reputation takes time. The less renowned estates depend on direct, and hard earned, sales. The majority of estates operate some sort of agritourism business. Accommodations almost ensure consumption of wine during the stay, more importantly they promise continued sales in following years to past guests. Bed and breakfasts are common – farmhouses, barns and more humble structures transformed into comfortable, often luxurious lodging, but generous hospitality is the key to sustained sales.

In Podere Terreno "every night is like a cruise ship with people from all different nationalities talking together." The lasting memory of these engagements, the word of mouth engendered when the tourists return home, and the quality of the wine, shifted Podere Terreno's reliance on direct sales to more than 50 percent in recurrent sales. Some estates offer only one or two rooms, others are small hotels. Cooking lessons, bus tours for tastings and so on are also used to bring the customers in. All these estates testify to the crucial impact the consortium has had in supporting Tuscan warmth.

Although pools and other comforts are offered with accommodations, few people don't venture out to other farms for wine and olive oil tastings, for hikes in the hills and among the churches and castles. The territory is of course ideally situated for day trips to Siena and Firenze. Throughout their stay the Black Rooster brand is ubiquitous, bought, used, and sold by restaurants, souvenir shops, and other businesses. The maps, signs and black rooster image enables tourists on bicycle, cars or buses to navigate the maze of hairpin roads and find wineries and sites along the way. It organizes meetings with journalists in the region and sponsors musical, gastronomic and other events. It also represents the territory at Vinitaly and other national and international wine shows, providing opportunities for winegrowers to attend or send bottles. The consortium's website allows buyers to research estates on the web and it runs ad campaigns and tasting tours in several countries.

Thus the marketing needs of the members are structurally differentiated, and the consortium is less or more relevant as the case may be, but there are other tensions that if not overcome, are managed within a common support for the territory. In their publications, the injustice of their usurped title to origin is used to build the mystique of the territory. Yet, many merchant-producers still use both the Chianti Classico and Chianti appellations despite the consortium's protestations against this theft of territorial birthright. All the cooperatives, Mellini, Badi a Coltibouno, Coli, Bartali, Piccini, and others, representing over half of Chianti Classico's production, produce wines under the Chianti label and/or one of the other Chianti appellations. The contradiction is emphatic for Badia a Coltibouno whose co-owner, Emanuela Stucchi Prinetti is a past president of the Consortium Marchio Storico and for Nunzio Carpuso the highly respected director of Mellini and past executive of both consortiums. The coops are strong supporters of the consortium and territory, but they serve winegrowers of different capacities within and outside Chianti Classico. They must preserve both the territorial and quality boundaries of the appellations.

All of these organizations are combining the effectiveness of offering a wide range of wines to their customers and the power of the Chianti name, the latter which can be used for a greater volume of wine than Chianti Classico.

The professed desire of the consortium to recapture exclusive use of Chianti is understandable because according to their research and the comments of several winegrowers, while Chianti brings instant recognition, consumers show little knowledge of distinction between Chianti and Chianti Classico. Yet the significance of market segmentation is demonstrated by others claiming that the distinctions are more or less recognized by customers. Despite the largest producers in the territory making Chianti, the majority of the bottlers, including the estate winegrowers, use IGT designations for their non-Chianti Classico, including attempts at supertuscans. A significant proportion of production may still be sold through grape or bulk sales into chianti or chianti classico blends. Chianti Classico, while the signature or dominant wine in volume for most estates, rarely makes up even half of total production. These alternative outlets represent a response to the quality standards of Chianti Classico on the one hand, and on the other, the need to produce a range of products for both sales and personal interest.

In 2001 Antinori and several large-scale Tuscan producers initiated a movement to create a Tuscan DOC. The objective was to capture an awareness of Tuscany among their international costumers that is even greater than Chianti and not limited to its territorial confines. Antinori claimed it would create a designation system similar to Bordeaux's.[31] The Chianti Classico consortium, its members and other territories defeated the attempt to create the appellation. They claimed Tuscany already had an IGT and the Tuscan DOC was an attempt to muddle the distinctiveness of its 39 DOCs, five DOCGs, and five IGTs. Emanuela Stucchi Prinetti is quoted as saying that the "whole manoeuvre has been designed by a handful of big names trying to make a quick buck without thinking of the consequences." Interestingly, Antinori, the most often cited consortium refusenik and large-scale regional producer-merchant, does not use the Chianti appellation. It prefers to use several IGT designations. Shortly before the consortium merger, Antinori withdrew its Villa Antinori brand from Chianti Classico to do so.

The struggle over the region's typicity and identity, to a certain extent, also resonates in the marketing differences of the estates and larger producers. The Chianti Classico differentiation is the key to the global market for the estates and at their scale of production they are likely to find customers. With larger production, more customers must be found and the inclusion of the international grapes allows the larger producers to broaden the appeal of their wines. The estates claim that it was the larger producers, with their volume-based voting power in the consortium that resulted in regulations allowing more Bordeaux grapes. Some also lament the fact that nobody will plant the traditional canaiolo because while adding fruit and finesse, it does not add body like the Bordeaux grapes. Ironically canaiolo makes a wine suspect in the new typicity.

On the other hand, attempts have been made to differentiate within Chianti Classico. In 2003, six estates tried to make an informal Super Chianti Classico name for themselves. This title was a play on more than the supertuscans. It was an attempt to claim a reputation as the region's best wines and best *terroir* and to establish a classification similar to Bordeaux. Barone Ricasoli, Castello di Ama, and Fattoria La Massa et al. hoped to be joined by other outstanding estates, but the initiative never gained momentum. Other internal differentiations have been considered but gone unrealized. The proposal to classify by elevation failed because of other *terroir* considerations. Generally, wine writers such as Belfrage, while declining to classify in a Bordeaux or Burgundy manner, use the communes to group wines. Some have even argued that regions such as Panzano deserve a higher status. The consortium and winegrowers are disinclined to introduce classifications they believe would be difficult to establish on a reasonable basis and that would be divisive.[32]

Without some system of internal differentiation, however, product-pricing differences in the market create consumer confusion and fury among estate producers, who rail that one territory should not produce such a disparity in prices. An estate owner, for example, had some explaining to do when a German customer saw much cheaper Chianti Classico for sale in a supermarket. Another estate was known to have dropped out of the *Consortium Marchio Storico* because of such differences. A large-scale merchant producer, however, was unrepentant about their price/quality ratio and was incredulous at some of the prices charged for wine, the production costs of which she knew very well. Yet, she admired those who were able to get high prices.

The consortium has tried to reduce the friction caused by price disparities by imposing a floor price on the sale of grapes and wine. Vineyard controls on reducing yields and on thinning bunches also force an increase in the price of Chianti Classico. These quality standards are also therefore quantity controls and they have been elevated after consortium gained complete control over the appellation. A recently introduced tracability system also goes some way to alleviate the price-based friction between the merchant blends and estates. If there are assurances that the quality of all Chianti Classico on the markets is within an acceptable range, it should be more acceptable to let the market do the pricing.

The great counterforce to the divisiveness created by disparate marketing approaches is the construction of the real and symbolic territory. Whether an image in distant markets or supporting direct sales to tourists, the territory supplies winegrowers with marketing at different levels of reputation. The consortium has been the primary protagonist of this territorialization. The creation of the consortium itself prevented the dilution of Chianti from a territorial definition to generic wine style. The Chianti Classico within Chianti compromise, while enlarging the area, still created a basis for territorialization. That the borders of the Chianti Classico appellation cut through two different Tuscan provinces and only includes parts of five of the territory's eight communes, yet is formally and informally recognized by the population, the industry and government, evinces the power of the consortium's territorialization.

The consortium has stamped its Black Rooster trademark on the territory through three generations of road signage; maps and memorabilia; tastings, concerts and other events; and of course on the bottles. The sophistication of territorial marketing jumped around the time that estates began to flourish. In 1975, for example, the consortium initiated a brand extension into olive oil.[33] The majority of winegrowers were producing olives and olive oils, but for bulk sales. With the consortium's Black Rooster logo they were able to add value through differentiation. Subsequently in 2001, a PDO designation was obtained and the consortium developed the expertise to certify these and products from other appellations. Moreover, mixing wine and olive oil increased the attractiveness to tourists, while giving expressive outlet to the estate owner: "you can criticize a Tuscan's wine but not his oil." Such activities, directly related to marketing, are important, but probably the consortium's actions to improve the actual territory more so.

Napoleone Franceschetti of Meleto initiated organized protection of the territory in 1970 by re-launching the ancient *Lega del Chianti*. Instead of military goals it fosters wine culture, rural activities and traditions, and the solidarity and friendship of the Chianti people. Much like a French *Confrerie*, the League draws its membership chiefly from the winegrowers, providing them with another venue to market and discuss collective concerns. The consortium formally began to organize for similar goals when it established the Foundation for the Oversight of the Chianti Classico Territory Onlus in 1991. The Foundation focuses on preserving the territory's cultural and environmental assets but its activities are mostly political. When the consortium believes the territory is under threat from incompatible development, such as industrial parks, roads, and garbage dumps it presses the issues with commune and provincial governments. For example it opposed a 24 hectares industrial site at Pianella and a 400 parking spot disco at Castellina. The Foundation and Consortium, are however, favourably disposed toward artisan industries.

Giovanni Ricasoli-Firidolfi, one of two winegrowing descendants of the Iron Baron and last President of the *Consortium Marchio Storico*, is a driving force in the foundation and shrewdly garnered financial and political support from the banks and other well connected sponsors. The Foundation also spearheaded conservation projects such as the several million dollar restoration of Radda's *Convento de Santa Maria del Prato* (a section of which was considered as the consortium's headquarters) and UNESCO recognition as a Chianti as a world heritage site. Although this restoration is supported by the winegrowers, Ricasoli-Firidolfi claims that Chianti needs the backing of the whole world. Chianti Classico's ecclesiastical relics, in particular, are in need of support as their adherents have deserted the territory and the relatively few replacements have not brought a devoted Catholicism to enliven the churches for tourists.

The winegrowers' activities create and preserve a territory that, while being part of Tuscany, is to a significant extent operated by outsiders for an external market. The winegrowers want to create an idealized environment to be enjoyed by tourists and symbolically by consumers in export markets. These activities draw them into a complex relationship with other claimants to the territory: inhabitants or would be inhabitants, business people, and the politicians that represent them. In the waning years of the *mezzadria*, the few former sharecroppers that remained and an influx of other small landholders looked to the communist party to support cooperative and mass quantity production of wine and urban and industrial development of the towns.[34] Chianti Classico's strategy of adding value through vertical and horizontal quality and the increase in tourists brought the community and politicians into greater alignment with consortium's territorial strategy.

The mayor of Greve obtained funds to support rural activities from the Tuscan government, while Radda's mayor was among the first in Italy to actively support agritourism and broker the national government's regulations. Through the 1990's Chianti Classico became a model of cooperative territorialization between the consortium and the communal, provincial and regional governments. The territory exemplified the convergence of agritourism, slow food, and respect for origin. Indeed the consortium has spearheaded a territorial practice that has been recognized by academics and bureaucrats. The EU has recognized such territorialism as a new source of regional development and for which it supplied funding and regulatory support. The mayors of the communes and the consortium joined together with unions, industry associations, and other stakeholders in the area to create a local area action group to secure funding from the EU's LEADER program for rural local development. They used the money to integrate and fund efforts on sustainability programmes, cultural events and restoration of rural buildings. These activities brought complementarity in activities, but did not resolve all differences. Tensions remained due to the wine territory's boundary excluding parts of some communes. More significantly, the interests of other stakeholders gained support from the mayors of the communes.

Respect for the territory's environment has been enhanced by the conversion of Chianti Classico to estate-based quality and away from the heavy use of chemicals in industrialized winegrowing. Winegrowers claim that with the zealous government protection of the forest, the limits on machinery use in the vineyards (for quality and soil maintenance reasons), and the need to emphasize *terroir* brings about improved environmental respect and performance. Taking advantage of this approach the consortium enlisted external support, especially through environmental NGOs such as Legambiente to argue against development. Still, the monocultural practices of the winegrowers leave them vulnerable on the sustainability front. The mayors used ICUN criticisms to argue for a vision of sustainable development that makes more room for more diverse land use.[35] As Ricasoli-Firidolfi's opinion of the damaging impacts of vineyard expansion suggests, many winegrowers realize the benefits of mixed development, and oppose irresponsible extension of vineyards. Not only can it take several years to get permission to build a winery, constructing any new buildings or renovating existing structures is strictly regulated. Buildings must retain the rural character of the *mezzadria*, even if that means a pig sty converted for luxurious agritourism must retain the tiny windows allocated to its original inhabitants.

Proprietors are, furthermore, acutely aware that restrictions on residential development force employees into long commutes and thereby increase their costs. Still these regulations are supported even if they impose significant inconveniences on management and operations. Indeed, although almost all estates have rooms for rent, many winegrowers are bothered by the impacts of an agritourism that is much more tourism than agriculture. At one time 80 percent of tourists to the region were staying on farms, now 90 percent stay in hotels, B&Bs and other accommodations not connected to agriculture. At the outset people would stay for a week, now only for two or three days. Buses, growing pressure on the roads, pollution, water supply problems accompany what some look upon as an increasingly mass tourism.

The need to preserve the territory for winegrowing and the need to protect the territory against winegrowing is the result of a rural renaissance with wine and gastronomy as heart and soul. They are experienced primarily through consumption but also through cooking classes and winery tours, with side visits to castles and churches to add extra purpose to the day. The winegrowers and other entrepreneurs put in long hours capturing their share of epicurean interest, but the movement has been supported by a number of organizations. The Slow Food movement has likely done most to popularize the gastronomic joys of *terroir* throughout Italy, Europe and to North America. Organizations such as *Città del Vino* support local government strategies for tourism, urban and regional planning, also combining to support territorial wine and local food industries, and to stage events. Small municipalities such as Radda and Greve have been helped to develop their own activities, collective efforts among he communes have been integrated, and *Città del Vino* helps the towns bring more to the table in negotiations or shared actions with the consortium. Tuscany proactively grasped the potential of territory-based gastronomy. It supports QC, funds marketing programs and wine fairs around the world, and brings in journalists. Perhaps its most remarkable activity is, working together with Slow Food, promoting organic and DOC products in schools, hospitals and other institutional settings.

*Enoteca-Italiana* organizes events, publishes materials related to the wine industry in Italy, and compiles and does market research. It works with ICE (Italian Export Commission) and Ministry of Agriculture in the support of sales abroad. An organization that is evocative of Tuscany and Chianti Classico's capture of enotourism is *Movimento Del Turismo Del Vino*. Donatella Cinelli-Colombini of Montalcino initiated the movement in realization of the importance of tourism, and provided the crucial insight that winegrowers would have to provide a warm reception. It was Tuscan women and their counterparts in other regions who brought about these changes and spread the movement throughout Italy. The operation of the wineries and estates were male dominated, "treating the cantina like the Vatican and thinking only about production."[36] Females had the detachment to realize that, for most tourists, the total vacation experience comes before wine.

The movement educated the wineries and estates, brought enotecas, restaurants, hotels and journalists into the organization and organized events. The two most important are the harvest welcome and the open cellars, the latter bringing in one million tourists. When the open cellars was initiated in the late 1990s only 25 winegrowers in Tuscany joined, but now over 200 participate.

The consortium participates in, or supports, many of the events and programmes developed by these other organizations and collaborates with other institutions to achieve economies of scale and scope. While it eschews involvement with the other Chianti territories, it works happily with the other high profile Tuscan sangiovese regions of Montalcino and Montepulciano on events, sharing information and as a united political front. Outside of the region it has a close collaboration with the Asti Consortium (*Consorzio di Tutela dell'Asti*) and the Union of Veneto Wine Consortiums (*l'Unione Consorzi Vini Veneti* DOC). They collaborate on the development of research, QC and promotional activities. Chianti Classico takes a leading role in Federdoc as well, particularly in the development of the quality control processes proffered for adoption throughout Italy's consortiums. Collaborative promotions with food product GIs, such as Parmesan have been successful as well. A lot of less-solicited support has come from the US. Frances Mayes' 1997 *Under the Tuscan Sun* and later Tuscan adorations played no small part in raising the US share of Chianti Classico's markets to 30 percent, a rise coinciding with an escalation in Chianti Classico's prices. The Wine Spectator's European editor's residence in Tuscany introduced a greater number of winegrowers to the US's more discriminating and moneyed consumers. In the same time period, however, German sales reversed in proportion to the American.

# Chapter 7
# An Invitation to Variety

Only variety can absorb variety.[1]

## Of Long Tails, Territories, and Differentiation

We expect a staggering amount of variety from a few industries. As of 2009 iTunes carried over 10 million songs and Amazon carried over six million English language books.[2] These are but a portion of the musical and linguistic selections that enrich our lives in multi-various and subtle ways. We search among different genres, artists, and styles, and the nuances are just as important as changes in kind. Plotlines originating in ancient Greece and the Bible are constantly reinterpreted and replayed in literature, theatre and dance; varying by theme, context, and literary devices according to author intent and actor expression. Despite the availability of superb recordings, Mozart continues to be played by millions of individuals and groups, ever more recordings generated, with each providing unique participatory and listening pleasure. The Beatles' *Yesterday* and Richard Perry's *Louie, Louie* have been covered by thousands of artists. Why shouldn't the appreciation of different genres and nuances be part of our diet?

Winegrowers invite consumers to discover and appreciate variety in an unprecedented diversity. They provide a wide range of grape varieties, grown in unique parcels of land; fermented, blended and lovingly matured in an infinite number of ways. Appreciation of this variety occurs at many levels, from the basic pleasures of grape characteristics or regional style to the nuances and complexities of *terroir* and technique. Millions of readers of wine guides, blogs, and magazines search out these not-so-subtle and subtle differences. The casual appreciation of wine variety, and more importantly, educating oneself to appreciate nuances, has become the benchmark for the evaluation of food and other products. This heightened appreciation is the cornerstone of increased value that has reinvigorated agricultural regions and recreated the relationship between country and city.

Winegrowers precociously developed a system capable of generating a variety that matches the latent demand for variety among consumers. We know that the demand for variety exists even if our understanding of how consumers select from conventional distribution outlets remains imprecise. It is increasingly clear that when the effects of assortment structuring and choice strategies are taken into consideration, consumers welcome variety and are not overwhelmed by too-much-choice.[3] The Internet has provided the technological means to serve the demand for variety in some areas and the resulting consumer surplus has been measured. Traditional bricks and mortar retailers can stock only a fraction of the tens of thousands or millions of items offered by their online competition.[4] This has dramatically changed the retail expression of Pareto's 80/20 rule. The top 20 percent of products no longer account for 80 percent of sales, rather small-quantity niche products, those not found in conventional stores, now comprise 30–40 percent of sales. Consumers have sought out or discovered the joys of variety when made available.

The long tail phenomenon, particularly exemplified in the publishing, music and electronics shows the promise for niche products, and may stimulate the development of new products, strategies and industrial structures. New industrial structures are especially evident in publishing and music where digitalization enables authors, musicians, and software designers the ability to self-publish and distribute online or with the assistance of intermediaries large and small. Alongside its Office™ and Explorer monopolies, Microsoft has always provided a platform for new application development and Apple and others with their "app stores" provide similar structures to gain network and other economies of scale and scope. Collaboration platforms are important to these new industrial structures, as the Internet has improved the quality and quantity of digital collaboration and facilitated the search for collaborators. Indeed, as exemplified by Linux and other open software, the creation of these industrial structures, is in itself a collaboration. The key to such collaborative structures is to reward those who contribute with more than learning or recognition. As with a winegrower territory, there is a need to create a common pool resource with legal and normative mechanisms providing returns to their efforts and which penalize free riding.[5]

Winegrowers have embraced the Internet for sales and communication purposes, while Internet merchants and auctioneers provide a wide selection of competing and complementary wines. American winegrowers, particularly, were excited in 2008 by the prospect of Amazon selling wine. Winegrowers, however, did not wait for the Internet before developing the long tail of demand and they have created sophisticated collaborative platforms that enable them to overcome the barriers of small-scale production and distribution. Winegrowers in all countries have and continue to develop territorial organizations that through collaboration achieve economies of scale and scope and through competition generate a profusion of variety.

The driver in the system is the recognition by winegrowers of their mutual need to differentiate their products by appellation and by estate. As a result, territories should be seen as more than the source of products of origin or quality products.[6] Indeed the use of the latter term, more intuitively associated with vertical quality is an unfortunate, if well meaning, misdirection of what agriculture can offer consumers. Consumer tastes and incomes vary greatly across the vertical and horizontal dimensions of variety and winegrowers can attempt to occupy a few of many niches. Likewise, territories are more than collectives, assemblies of like-minded producers agreeing to produce similar goods[7] or constructing unique places. In the process of creating their territories, winegrowers create wonderful landscapes and communities. Although for many these means becomes an end,[8] without consistent attention to differentiation and to market creation, such efforts can lose their foundation. A focus on differentiation, and its ramifications can bridge the divide between place-based and market-based views of what quality is and provide a stronger vision of what alternative food systems can be.[9] The market is for variety. The winegrower, with the support of their territory, meets that need through differentiation.

## The Territory: A Framework for Diversity

> The issue for wine territories in France is to be professional. (Hubert De Bouard, Past President *Syndicat Viticole de Saint-Émilion*)

The value in comparing Bordeaux, Napa, and Chianti is to produce a framework for organizational structure that identifies key collective action issues, and the mechanisms used to enhance cooperation among the territory's members and within the distribution chain. The framework has to incorporate a high degree of flexibility because while the three territories have shared common challenges, they have experienced unique trials. Similarly, responses have been akin and innovative. Most importantly, while an organization must remain vigilant to changes in market and political governance, it must also respond to the evolution and diversification of its winegrower members. The framework is built on the premises of common pool resources described earlier, including: the need for rules to reduce free riding and information asymmetries and increase transparency; monitoring and enforcement of rules; repeated interactions, credible commitments and other means to build social capital; the effective use of third parties; and effective participation in decision-making.[10]

Effective and efficient systems for competitive differentiation – in marketing, appellations, and chateau identification (classifications) have to be built into the framework to make it the basis for a differentiated common pool resource.

### Democracy, Legitimacy, and Efficiency

*Benefits of self-governance* The core of a territory is voluntary self-governance imposed by winegrowers on themselves. The benefits that can accrue from an appellation depend on how it is consistently supported and its integrity defended. Anyone within the boundaries of an appellation can use it by following basic regulations or paying minor fees, but that does not mean they will get the most out of it. With substantive organization, participation and investments, local winegrowers generate opportunities to learn how to protect their crops from diseases, obtain the discipline of quality control, enjoy forums for information exchange, for marketing opportunities, and other means to support individual and collective reputation. If winegrowers attempt to benefit from such activities without participation or payment, self-governance must be prepared to deal with free-riding and create a transparent, legitimate and efficient governance system.

*Democratic structure*    The democratic structure of self-governance varied subtly among the three territories. All elected presidents and executives, but whereas Bordeaux and Chianti Classico elected these representatives for long terms or allowed repeated mandates, Napa elects a new president and executive every year. Where the Europeans have sought, and especially in the case of Saint-Émilion and Blaye found excellent extended leadership, the Americans insist on new blood. In all these cases, including Napa, repeated election to the executive was a prerequisite for presidency, the president and potential candidates had to be respected winegrowers, and the position alternates between different statuses within the territory. In Saint-Émilion, for example, Jean-Francois Quenin, owner of the Grand Cru Chateau de Pressac, became president in 2008, after eight years' leadership by Hubert de Bouard of Premier Grand Cru Classé Chateau Angelus, who had taken over from the Grand Cru and Saint-Émilion producer Jacques Bertrand of Chateau Carteau Côte Daugay. In Napa, the presidency continually revolves through small, medium and corporate winegrowers, a manager from the latter usually taking the turn.

*Legitimacy*    Surpassing the executive in power, winegrowers in all territories vote to decide important matters. When this power is superseded by presidential presumption a crisis in self-governance results, such as occurred in Napa with the winery designation ordinance. The construction of legitimacy requires more than an occasional vote from the plurality, however, because of the different strengths, needs and groups of estates. And while territorial governance is disposed to the needs of estates, it has opened to equal membership of cooperative members and of corporations. The later introduces a temptation of larger landowners and producers to ask for a greater voice. To ensure minority positions are not overwhelmed, different status estates, young winegrowers, sub-territories and other groups are accommodated with reserved places on executives, commissions and voting procedures.

*Executive and administrative efficiency*    Perhaps the greatest benefit of territorial executive governance is the hands-on leadership that winegrowers bring to leadership. They are all entrepreneurs with extensive experience and knowledge of the business. They devise the strategies and policies of the associations, taking direct charge of commissions on promotion, research, quality control and so on. All the territories in this study were supported by administrative structures that generally mirrored corporate counterparts in terms of departmental divisions of labour. Promotion activities, in particular, followed many of the same precepts as the corporate example. On the other hand, the directors and employees were acutely aware of the challenges and the need to deal with their many estate and cooperative employers. It could easily be said that territorial managers feel the governance of their employers more than corporate managers feel the impact of their shareholders. Other outstanding features of the administrations are their small size and costs compared to the services they deliver and the production volumes, size and complexity of the territories. Although not completely unanimous or without gripes, winegrowers expressed considerable respect for Linda Rieff (Napa) Giuseppe Liberatore, Christophe Chateau, Jean Lissague, Roland Feredj, and Nadine Couraud.

*The freerider problem*   Perhaps the most contentious actions that can be taken by an association is to establish collective controls on winegrowers activities and to extract dues for collective action. Making such decisions is an expression of collective will because regulations will impose compliance costs directly on winegrowers and also cost the association monitoring and enforcement costs. Failure to establish such mechanisms, however, may impose greater costs by damaging or not building the reputation of the territory to its potential.

A good deal of investigation and open discussion is necessary to gain acceptance on an initiative. Even if these common purposes can be agreed to and formally enacted, winegrower associations have to prevent freerider activities: to ensure the success of the action taken, to preserve the reputation of the territory, and moreover to appear to be fair to all winegrowers. In Chianti Classico and Bordeaux, the mechanisms to deal with freeriding were institutionalized with the quality controls meant to upgrade hundreds of winegrower to a minimum level. Monitoring and enforcement in these systems was never perfect and associations are under pressure to constantly enhance legitimacy of governance. Moreover, they have to raise funds for marketing and link expenditures to winegrower incomes. R&D also requires collective funding and the potential to improve individual results. Contributions to territorial development and even sharing in the reduction of acreage are issues vexed by the freerider issue.

The mandate to govern the activities of all winegrowers that use the appellation in Bordeaux and Chianti Classico allows a high level of control over freeriding. Napa, of course, within the AVA system doesn't benefit from this control, nor did its evolution of quality indicate the need for such a mandate. From a handful of estates in the 1960s, an incremental development of vertical and horizontal quality combined with a constantly growing reputation to bring investments in talent and equipment that, for the most part obviated the need for quality control. The territory did, however, seek control over winery activities and development, needed collective funding of R&D and pest control and for worker housing and health care. Lacking a self-governance mandate and dealing with the issues that fell within disparate governance spheres, the winegrowers turned to county, state and national government institutions for governance over freeriding. The practices used to control freeriding provide much of the substance of all governance activities.

*Regulatory Governance*

*Scope of territorial mandate*   Territories live in a circular relationship with governments. From the outset, they owe their existence, at least partially to a government mandate, but winegrowers have called upon governments to create these mandates. As time passes territories must continually lobby for policy changes and must be prepared for the government to bring in changes that affect them directly or indirectly. The key policy and regulatory set is that of the appellation system with its associated estate, wine proportion, and labelling rules. These rules vary by nation and, as they have significant ramifications for winegrower viability, continue to be issues of discussion and change.

Two other important structural variations are whether or not a voluntary association must support an appellation and whether that association can impose standards or other expectations on its members. On the surface clear differences distinguish the French model's requirement for an association and its strong mandate for territorial governance from the US territory's freedom from associational and behavioural control requirements. The NVV's pre-appellation existence and its struggle to obtain laws to govern use of its name and activities of winegrowers demonstrates, however, that aspects of a regulatory mandate for control is desired by territories that don't have such a mandate. Chianti Classico's long struggle to get the capacity for self-governance and the rapid implementation of such controls when finally mandated furthermore demonstrates the utility of a sustained association. Undeniably, the conditions for and types of associational powers vary, but it seems a government mandate for control is universally desired and has been employed with substantial effect.

*Control, scope, and advocacy of regulatory body*     Who controls the regulators differs significantly. In France winegrowers control the national committee of the AOC and of the INAO. Logically and in practice this control is more supportive of the territories, providing winegrower associations with a clear mission of territorial definition and support for *terroir*-based production. Winegrowers and association leaders on the national committees provide practical perspectives on the needs of producers and their territories. These interests are parlayed to the ministry of agriculture that, although limited to veto power, still provides an institutionalized and powerful voice to other government departments and politicians. Winegrowers at the territorial level, occasionally chafe at CNAOC and INAO oversight, but overwhelmingly support a system operated by winegrowers for their interest.

The Italian national committee for DOCs appears to have similar powers to the French, but its role is more of an advisor to the government. The bureaucrats and politicians designed the appellation system and determined who would get DOCGs rather than the winegrowers. Furthermore, committee membership is broader and with the winegrowers priorities subject to the interests of industrial producers, merchants, cooperatives and bulk producers. As a consequence new DOCG regulations still accommodate industrial volume producers and alienate estate winegrowers.[11] Fortunately, in 2005 after a long struggle Italy's estate-focused *consorzios* were able to convince the politicians and get the legal mandate for control of their territories. In the US, the then BATF held stakeholder hearings to help devise the appellation system, but it was not devised by or for winegrowers, nor do they have much say in the running of the system, and who for example, can establish an AVA. Petitions are made to the now TTB and the bureaucrats decide based on their criteria.

Winegrower control over the appellation system is important to ensure that new appellations meet standards for horizontal differentiation of territories and for assurance of vertical differentiation (or quality). At least that is the perspective of the Europeans and many writers who look to appellation systems to express a logic of quality. The American's eschew formal indication of vertical differentiation through AVA designation. Perhaps the focus on horizontal differentiation is the most important role and bureaucrats can arbitrate sufficiently. However, who controls the appellation systems is also important for lobbying for territorial rights within society. French territories are federated and through their national committee advocate for respect for collective and territorial intellectual property right protection, for other policies, and resources. The INAO also pursues trademark infringement, nationally and internationally, relieving winegrowers of an infeasible cost. Italian territories, through the national committee and independently argue for similar rights and have been particularly vocal in lobbying the EU for greater protection of origins and trade practices used by territories.

The TTB plays no advocacy role whatsoever and as demonstrated in the Calistoga AVA-brand squabble can be dismissive to winegrower concerns. The need for Napa, and subsequently Sonoma, and then other territories to obtain greater protection of their trademarks suggests a common need for self-governance, not only at the territory level, but also of the appellation system. The US relies on a rather loose coalition of territories and lobbying groups to present the winegrower position at the state or federal level. The most important industry lobbyist, the Wine Institute, has only recently joined the coalition, overcoming its favouritism towards industrial production and antagonism with the Family Winemakers of California. A new organization, Appellation America, not only advocates greater legal recognition of appellations, but also promotes them to the public. Direct promotion is something that the French and Italian appellation systems do not engage in, although their governments are much more supportive. Their committees focus on regulations and legitimization of the concept, increasingly so.

In France, for example, the success of the AOC system encouraged the government to expand its usage and the INAO's governance to dairy and other agricultural products. This policy was couched in attempts to revitalize rural areas and evolved after the creation of a European system of PDO's and PGI's in 1992. This advanced the INAO's interest in legal recognition to the AOC concept, but the system no longer focused on winegrower interests. They soon shared a revolving leadership of the CNAOC. New systems of governance were introduced, most notably committing the INAO to control products not linked with their origins. PGI's, for example, are loosely linked to geography and standardized through a certification only requiring two characteristics to set it apart from conventional products. Similarly, the Red Label for meat is provided based on achieving a standard of vertical quality. Both provide differentiation, but for a standard much more akin to a mass-volume brand.[12] The INAO is also entrusted with two organic and one traditional specialty certifications, which again are a standard not specifically linked to place or differentiation by producer.

The drift toward standardized certifications is in line with the growing influence of quality control systems, and in 2007 these were incorporated into the reform of the *syndicat's* locally managed certification of winegrowers for the appellation. At the same time, the National Committee (CNAOC) was transformed into a wider body of 24 representatives drawn from the wine and brandy, dairy, agriculture-food, and PGI committees. The wine and brandy committee remains the first among equals, but its committee size was reduced from 83 to 50. Thus, while the CNAOC-INAO is undoubtedly increasing recognition of appellations, one can question whether the dynamic of differentiation within a territory is being compromised.

*Ancillary governance*   Several other governmental and social forces, outside of appellation regulations, exert indirect and powerful influences on territorial governance. Indeed they demand a substantial collective response from not only the territorial association, but the coalitions of other territories and industry stakeholders. As discussed in the chapters, these influences are exemplified by health and environmental issues. In France, the stronger territories, the winegrower and merchant federations, and their political allies, have stepped in to counter an anti-alcohol movement supported by a Health ministry concerned about physiological impacts and road deaths. The coalition attempts to separate wine's influence on a healthy lifestyle from the generally deleterious affects of alcohol. American winegrowers, while concerned about neo-prohibitionism, established proxy organizations to engage in a legal and publicity battle against the cynical use of State alcohol regulations by wholesalers. Environmental issues impinge on several levels. Territorial associations need to negotiate regional regulations and subsidies that deal with effluents, pesticides, and biodiversity, while national, Federal/Community regulations govern greenhouse gas and bottle recycling concerns. Environmental NGOs and society generally put significant pressure on highly visible industry.

Responses to health and environment issues are collective reputation issues, open to not only significant freeriding, but also to the necessity to respect the varying capacities for response of different winegrowers and the threat of the collective invading niches developed by winegrowers (e.g. organic practices). The response that territories and their coalitions devise for ancillary governance, thus, while recognizing the multi-faceted nature of each challenge, must also keep the complexity of its own members' perspectives forefront when making policies. National context is important, as illustrated by comparing state support for the healthcare and pensions of employees of Italian and French winegrowers with their American counterparts. Napa has to fund and partially organize these services, along with needs such as worker housing. To do so it has devised means, such as the auction and legalized mandatory assessments, to overcome the tendency to freeride and to legitimize the response to winegrowers.

Other issues such as the pervasive regulatory red tape that confronts alcohol production and sales is lobbied against and to dealt with directly to control transaction costs. Associations offer their members legal repositories and deciphering of regulations and interact with authorities to standardize and integrate reporting forms. Perhaps the most general need is to push the authorities of ancillary governance to recognize issues related to winegrower scale and continuity, particularly in regard to tax burdens. The Americans have secured tax concessions for small winegrowers and the French secured relief from family succession taxes.

## Boundaries and Signals

*Exclusion and inclusion* Appellations arise when winegrowers from a particular geographic area obtain the right to exclude others from using its name. Among the three nations, exclusion is done on similar standards in regard to an area's biogeoclimatalogical features and human history. In order to minimize the inevitable conflicts of interest as to who gets included and excluded, a third party arbitrator is always used. The legal definitions of appellations divide in a Europe-New World fashion when continued exclusion is necessary to sustain an appellation, especially as it takes on value. Territorialization to reinforce exclusion (and promote the appellation) then rejoins the two types. The small size of most winegrowers requires collective defence against appropriation of the appellation, and further underscores the necessity for some sort of sustained organization. Napa illustrates. It was granted an appellation under the AVA regulations but has had to use existing laws, sponsor new laws and organize coalitions to defend its monopoly on case-by-case and systemic bases. Newer territories are learning to take these steps.

The winegrower-government interaction that creates the appellation monopoly[13] is important, as is the need to use geographical and other characteristics to structure internal differentiation within the boundaries. The main tools for this differentiation are *terroir*, and its classification and sub-appellation-territory derivatives. They are organized, funded, and mediated through collective self-governance. The concept of *terroir* provides each winegrower with the means to differentiate. Besides cultural and ideological support, the associations provide R&D, training, and build their promotional efforts around this concept. Bordeaux's classifications are inherently self-governing, either in conscious acceptance of the Médoc's frozen and successful classification or a Saint-Émilion style periodically reviewed system. The legal challenges and defences to Saint-Émilion's classification (and others) exemplifies the difficulties of self-governance, and that for better or worse, even in France, territorial self-governance has difficulty establishing its autonomy and legitimacy within society.

Although *terroir*-based, the classification systems are an assortment tool for the vast amount of variety in each territories, supporting the division of Bordeaux into many territories and appellations. Thus far, Napa's winegrowers haven't needed a classification to distinguish among the 10–30 competitors within a sub-appellation. On the other hand, Napa precluded potential Balkanization of the main reputation by legally protecting the primacy of the mother appellation over its children. The Bordeaux name seems not to require such protection, but winegrowers in Saint-Émilion are confident they could thrive without the Bordeaux umbrella, and could do without the broad-brush criticisms underperformers have inflicted on the larger territory's reputation. Chianti Classico chose the independent route, although with a stronger rationale for not confederating. Yet one has to wonder if an earlier granting of self-governance would have allowed the geographic and qualitative differentiation that could have produced a broader and stronger Chianti territory – one that could combine the widespread recognition of the name with an ability to define leaders, capitalizing on scale and assortment of variety. Chianti Classico is not afraid to work with other leading territories, but believes it is unnecessary or too problematic from a governance and legitimacy stance to define sub-regions or classifications.

Territorial exclusion and inclusion exist in complex relationship with brand power. Big brands exist in all the territories and account for half or more of all volume sold. *Négociants* are the traditional blenders in Bordeaux, but famous chateaux, large companies, and cooperatives also make Bordeaux and other territorial brands. In Napa, big wineries like Mondavi, Martini, BV etc. dominate, but other internal and external firms, some recently calling themselves *négociants* also blend in volume. Chianti Classico has a number of large wineries buying grapes and wine for large volume sales. On the one hand brands represent the territory at the largest scales of *terroir*; they represent the territory through a more extensive form of promotion, and provide an affordable introduction to the territory's style. They integrate the territory in a palpable production system and may well have been the origin of a territory's cohesiveness.

On the other hand, volume production contrasts with the premise of estate wineries. The quality and over production problems that damage the territory's reputation are usually traced back to brand wines, moreover it is the temptation to exploit the appellation (or a brand related to the appellation) for volume production that prompts firms to blend in imported grapes, use external grapes at a winery in the territory or use the name on wine made elsewhere with other grapes. Complicating matters is the fact that owners of both estates and volume brands want to expand their brands while protecting their estates. These problems have bedevilled Chianti Classico, risen recurrently in Bordeaux, and afflicted Napa. Several mechanisms of territorial governance have been developed to overcome this dissonance, quality and quantity controls on the production side, with the support of legal protection of appellation content and government enforcement. Moreover, the associations, interactions among winegrowers of all types, and commitment to the territory have helped to overcome some differences.

*Grapes and Styles*

Next to a territory's name, its wine style is the most identifiable signal it can send to market. The Bordeaux blend evokes the power of a style resoundingly. Chianti Classico's championing of sangiovese and Napa's de facto concentration on cabernet sauvignon demonstrate further the powerful linkage between market recognition and style. The message is best understood as a consistent and coherent signal, that is, most winegrowers in the territory make similar style wines. That is not to say that the styles enforce conformity. The Bordeaux blend varies from winegrower to winegrower and from territory to territory, most obviously in the cleavage between a cabernet sauvignon dominated left bank and a merlot dominated right bank. Most Napa cabernets are closer to a Bordeaux blend, and its variations, than a straight varietal. Chianti Classico's sangiovese has only recently expelled white grapes and allowed for a pure varietal, but most wines remain a blend. The creativity of selecting different grapes to plant in differing *terroir* remains a feature of territorial style.

Yet there is also a need for diversity of kind within the territory and within a winegrower's product range, at least at the scale of the broader territory. The dry and sweet whites derived from semillion and sauvignon blanc are Bordeaux blends as well. They make dinners in the territory, perhaps at chateaux, much more interesting. Napa restaurants can serve locally grown wines that will suit virtually any meal they can serve, while the estates can usually offer their club members a suite of different varietals and blends. Chianti Classico is focused on sangiovese dominated red wines, but you can still round off a repast with vin santo. From both winegrower and territory perspectives diversity of styles provides a better range of products, increases competition and provides an outlet for creativity. Most importantly, creative winegrowers want to play with several varieties.

The evolution of cabernet to pre-eminence in Napa illustrates that it is not necessary to stipulate the use of a style. However, that evolution occurred from a low level of cabernet planting and was accompanied by the influx of new ideas, money and market awareness. Bordeaux's regulations existed from 1935, and the vast majority of winegrowers adopted the style as they moved out of white wine production. Many territories now possess under utilized appellations that were created because winegrowers had planted poor varieties, but couldn't convert the vineyards at that time. Chianti Classico is the only territory that has asked pre-existing members to change their grapes varieties. It had to do so because the evolution of its wine style was suppressed by regulations for decades. Yet while the *consorzio* attempted to telescope style and regulatory evolution into a couple of decades they still had to provide winegrowers with reasonable timelines, allow them to experiment with new varieties and proportions and provide new appellations as outlets to market the grape varieties no longer allowed in Chianti Classico. Still, it was a messy, contentious process. The three territories demonstrate the value of a coherent wine signal, but as their styles evolved over long periods, it would be wise of newer territories to be wary of thinking they can replace time with regulation.

## Distribution Chain

*Bottlenecks and lock-ins*    If winegrowers can produce and consumers can absorb variety, their happy marriage remains hindered by three bottlenecks. The first is between grapegrower and winery (whether private or cooperative). The grapegrower can express *terroir*, capturing the nuances of her land with viticultural techniques and choice of varieties, rootstocks and so on. But, although the rhetoric is that great wines depend on the vineyard, grapegrowers struggle to capture the value of land and effort. The failure of the commodity chain to reward or even care about such differentiation thus turns many grapegrowers to winemaking. Some increase in value can be captured through making wine for bulk sales (and costs of marketing etc. avoided), but this short reprieve from marketing a perishable product limits *terroir*-based differentiation. Estate winegrowing captures that potential, but opens the challenge of branding and marketing. Estate winegrowers have to get through the final bottleneck, a distribution system constricting with the consolidation of wholesalers and retailers at the same time variety is proliferating.

The bottlenecks, counter-intuitively, could be explained by technical and managerial efficiencies or the elimination of double-margins. Consolidation of distribution, especially, points to wringing transaction costs and variety out from the system. The evidence, moreover, points to the compounding effect of the lock-in[14] of conventional practices among the transaction partners in the value chain, within the institutions that govern the value chain, and at different levels. A few examples clarify these points.

Beckstoffer and others lead a struggle against what was both an economic and social relegation of Napa grapegrowing. In Bordeaux the rationale of bulk winegrowers was to make the vines piss in order to maximize their returns, and for some bird hunting was a greater priority then tending the vines.[15] The mixed untrained vines (promiscuous) in Chianti Classico's vineyards is a legacy of the sharecropping past. The CIVB was created to mediate the bulk winegrower-merchant relationship and its most obvious tools remain dedicated to that role. Grapegrowers and vintners in Napa remain in an associational schism produced by historic antagonisms. Chianti Classico's evolution was locked-in by an externally imposed political process of vested interest competition and with limited understanding of differentiation. Napa, and US winegrowers generally, remain constrained by the compromises used to end prohibition.

*Institutional innovation*    The bottlenecks and lock-ins have been partially relieved by a number of institutional innovations. The most fundamental is bringing transparency to the distribution chain. The registration of transactions between bulk producers and merchants, and the publication of average prices remains the primary tool of the CIVB, and provides Bordeaux's winegrower and merchant sides with a basis of understanding. The territorial and appellation analysis of these transactions also supports competition and differentiation in the broader territory. Napa's instigation of the crush report has brought much fairer transactions to California, and perhaps its segmentation by counties could be extended to sub-appellations. Chianti Classico attempted to simplify transactions by setting a floor price.

Quality contracts between wine/grapegrower and buyer are used extensively in many configurations in Napa and Bordeaux. Although bottle pricing hasn't been widely adopted, vineyard designates have given grapegrowers a powerful tool to capture the value of *terroir* and viticulture. Grapegrowers profiting from vineyard designates remain primarily a Napa/US story, but Bordeaux and Chianti Classico's cooperatives reward differentiation among their growers, for the health and quality of their grapes and for *terroir*. Many cooperatives make individual or small group brands, stressing their *terroir*. In Bordeaux, *vigneron* and *négociant* share the value of differentiation with merchant bottlings of individual chateaux – a creative response to the constraints imposed by estate bottling regulations and the value consumers attach to estates.

Bordeaux has three institutions unique among the territories. Although Chianti Classico and Napa have merchant-blenders and large integrated producers, only Bordeaux has a corps of merchants dedicated to selling the products of many competing chateaux. Each of these merchants offers their customers a range of Bordeaux, often including their own label, and occasionally rounding out their offer with products from other regions. The CIVB has diversified from its initial bulk mediation role to support estate sales in supermarkets and trade fairs. Third, only Bordeaux has successfully created a futures market that reinforces differentiation early in the distribution chain and enables winegrowers to capture some of the speculation value generated by their wines. Orchestrated by the Union of Grand Cru, the *en primeur* market and its compliment *la place* have a profound effect on Bordeaux's locomotives, and winegrower associations draw an extensive impact back to their territories and chateaux. Napa has tried to replicate some of this success with its two auctions, but not only does Bordeaux have tremendous first mover advantage, Napa lacks the corps of merchants that balance the relationship. Burgundy and the Rhone have smaller corps, reflecting a size difference, but neither possesses a market and merchants so dedicated to dealing in variety.

The overwhelming strategy of estates in Chianti Classico has been to sell direct. The most renowned wines find agents in export markets. The less renowned estates focus on direct sales at the door and to customers who keep the memory alive. In neither case is there an institution focused on mediation of transactions, although the *consorzio's* quality control enhances relations.[16] Napa's smaller wineries and estates focus on direct sales as well – at the door, wine clubs and Internet sales. With great expectations for direct sales beyond the cellar door, winegrowers have had to establish and support the associations of Free the Grapes and the Coalition for Free Trade. The majority of Bordeaux's petits chateaux do not feel the *en primeur* effect greatly and must market extensively through Europe and internationally. Vigneron Independents supports the efforts of its members with wine fairs, promotions and other activities.

How were these innovations achieved? Some such as quality contracts and vineyard designates have arisen as entrepreneurial initiatives diffused, been adopted and adapted according to the needs of the transaction partners. Often with diffusion, the standards and expectations associated with the innovation becomes a business norm. Associations are another source. Winegrowers and merchants have used existing associations to discuss issues and devise solutions, some they implement themselves, while others require a government mandate (e.g. Bordeaux's transaction register) or laws (e.g. Napa's crush report). Government lobbying and coalition building may be necessary. Most importantly these innovations are based on constructing viable mechanisms that can produce value; are evident to both sides; or balance the terms of trade to allow both sides to generate value. The institutions responsible for introducing and maintaining the mechanisms don't have to be territorial or even an association. The Vigneron Independants is a national organization and many in Napa get their information from consultants such as MKF or Silicon Valley Bank (which often works with the NVV).

## Primacy of Reputation

The most important distribution chain tool is the territorial reputation. The majority of small winegrowers don't have the marketing capacity to gain sufficient attention and achieve their goals for differentiation and market power. The territory's reputation provides leverage in all stages of the distribution chain. It draws merchants to a territory and sets a floor price. Retailers organize purchases and shelves by territory. Territories draw tourists for direct sales and are the brand name most familiar to consumers in distant markets. For the majority of small producers the appellation is more important than their own brand. That said winegrowers want to stand out within their appellation or surpass its reputation. Territories, must therefore, balance this need as they build their reputations whether it be in the classifications, promotions or quality control.

## Economies of Scope and Scale

Integrating collective and individual reputation building is the primary challenge in creating viable economies of scope and scale. Viability and legitimacy requires effectiveness, occasional monitoring and enforcement, tolerance for differences in abilities, and fairness in resource allocation. Territories stress economies of scope over scale, but there are a few activities where size matters.

*Promotion*   All the territories try to brand themselves, with identifiable logos, putting their logo on estate labels, in interactions with the media and in their publications, and explaining why their territory produces the best wine in the world. Only Bordeaux, however, engages in significant generic advertising. The vast majority of promotion stresses *terroir* diversity, the differences among the estates, and makes navigating the territory legible on the ground and in cyberspace.

Connecting territorial and individual promotion outcomes is important for effect and legitimacy. Associations spend most of their budgets on promotion, the majority of money coming from the membership. Yet it is hard for members to tie territorial promotion directly to the sales of each winegrower's brands. Thus websites have links to all producers and feature a diversity of wines at tastings, trade fairs and so on. To ensure fairness there must be a fair and transparent selection of wines. The Napa auction exemplifies how a balance is struck. The high profile event brings the territory great exposure and provides a vehicle for each winegrower to stand out. The donation, however, is expensive for small winegrowers, especially if they want to make a splash.

*R&D and training* Continuous training and adaptation of innovations from around the globe and a local capacity to innovate are crucial to vertical and horizontal differentiation of territory and winegrowers. Local institutions of training and R&D are the backbone of these efforts. Bordeaux and Napa are rich in all levels of institutions, from training pruners to PhDs researching plant diseases or marketing. Chianti Classico enjoys less training support and there is no local oenology faculty, but it does work closely with universities. Most institutions are largely government supported, but there are mechanisms for winegrower funding through associations or in California through donations. These mechanisms provide winegrowers a say in research and training programs.

The associations develop close associations with these institutions to bring the research and training into the territories, to adapt to local *terroir* and practices and to reinvigorate them. Surveys and experiments are conducted, training programmes and seminars held. Perhaps the most important approach is to create a culture of improvement and diffusion. The associations form committees to discuss technical and marketing issues and to develop programmes to deal with them. The sharing of information among leading winegrowers and diffusion through the territory on an informal basis, and by watching neighbours, extends the culture of improvement. Notably the culture of cooperation was shared among these three territories and many others through the sabbaticals and discussions of winegrowers and employees.

*Production and quality control* Production and quality controls support the most important economy of scale, they protect the collective reputation. Collective quality control also provides an economy of scale by presenting winegrowers with an affordable external check. These controls can also be divisive. Limiting production or denying the appellation seriously impacts a winegrower's income, but allowing underperformers and freeriders to escape control damages the interests of all. Associations have to proceed with agreed-upon standards, transparent assessments, support for improvement, third-party support and impartiality, and funding. Even with such parameters, a perfect system is unlikely to be developed because of the need for flexibility in differentiation and innovation.

No territory has developed an effective means to limit the expansion of vineyards, especially without constraining the ability to expand on good land. Of the three, regional authorities provide closest guard of vineyard expansion for Chianti Classico. Bordeaux still struggles with the rampant expansion of the 1970s and 1980s, while Napa perceives the first hints of possible over-expansion. European planting rights and vine-pulling schemes, and resulting national politics complicate self-governance of this issue. Generous funding, the hiring of star winemakers and a culture of technological excellence has thus far precluded the use of external quality control in Napa. Bordeaux's territories have recently revamped their quality control systems in response to internal and external demands for more rigour. Downstream monitoring, however, is conducted by the CIVB, but not completely embraced by the territories. Since gaining authority over all who use the appellation, Chianti Classico has introduced the most rigorous monitoring in the vineyard, winery and tracability of the product in markets. Not to be forgotten is Saint-Émilion's creative combination of quality control and classification to escalate standards and as a competitive tool.

*Politics*   One of the greatest advantages of corporations is to leverage financial clout, employment, and lobbying resources in efforts to win regulatory and political advantages. Industry associations provide winegrowers with a unified voice, for without such scale, the voices of thousands of petty producers would not be heard. Associations and federations or coalitions of associations are critical to help winegrowers build a political voice, and on occasion, counter lobbying by large players in the industry. Coalitions may be formed with aligned organizations or those outside the industry to obtain support on environmental, development or cultural issues. Territorial associations are usually sufficient to deal with municipal and regional governments, tourist boards and so on, but not always. Since local and national regulations usually intertwine, territorial associations call on the support of their federations to bring in added weight. They also depend on them to carry their arguments to national and international politicians and bureaucrats.

A critical issue for winegrowers on all continents is international respect for the concept of appellations and origins. Primarily, that involves overcoming regional and national resistance to push national or EU governments to lobby for respect at the TRIPs conference. Similar protections are sought for related issues such as the use of viticulture or viniculture practices developed in the territory and terms such as rosé or riserva. These practices provide a means of differentiation, the utility of which has been degraded by pervasive use and without being backed by any qualitative difference. International cooperation among associations to promote the concept of origins to a broader audience have recently been initiated, such as Joint Declaration to Protect Wine Place & Origin and Great Wine Capitals, but these remain consortia of independent wine regions rather than carrying the force of federations. Perhaps greater popularization of *terroir* and place has been achieved by the slow food and other movements.

The tools used by the associations and federations are similar, but adapted to scale. They are based on a dossier of how important winegrowing is in terms of direct and indirect employment, contribution to GDP, role in tourism, sustaining local environment, and enhancing quality of life. The dossier reminds politicians and planners of winegrowing's relevance within a rapidly changing economy and urban system. The electoral power of winegrowers and employees is significant. In rural areas they make up a substantial proportion of voters and can elect representatives dedicated to their cause. Napa and Bordeaux have been successful at this game at both the state and national level. There are differences in the political structure, however. Bordeaux and Chianti Classico enjoy federations among winegrowers, and also for Bordeaux among merchants and inter-professional organizations (i.e. counterparts of the CIVB). Napa enjoys integration of the grapegrowers with the powerful Farm Bureau, but the NVV is not federated with other territories. Indeed the Wine Institute, with its tension between estate and industrial perspectives, remains the key industry presence at the state and national level.

*Management*    Perhaps the most important economy is the delivery of scale and scope advantages with a minimal and efficient managerial structure. As an indicator of this efficiency, the size of all administrative structures is disproportionate to the volumes of production and employment in the territories. The winegrowers are not burdened with bureaucratic costs and can focus on their own operations.

Association directors and other key personnel are retained and develop a great deal of institutional and territorial knowledge. Most of the actual marketing design or analysis, research and so on, however, is outsourced. Very often the outsourcing is on a minimal cost basis, using a sympathetic professor for geological analysis or a consultant that provides services to members. The support of FEDERDOC, the INAO and TTB on the establishment and operation of territories reduces costs, as do alliances with governments, trade fairs, chambers of commerce, and tour companies. For the most part, the administrations focus on their key missions. Only the *Consorzio Chianti classico* has diversified into activities, such as representing olive oil producers and offering laboratory services to other territories and industries.

Voluntarism at executive and implementation levels is the key to territorial efficiency. The executives are accomplished winegrowers, successful in international competition and know the challenges of their industry. These executives have worked their way up through associations committees, developing and exhibiting leadership abilities. Certainly politics is involved, but high levels of expertise among members expose inabilities of policy formation. The most fundamental activities of associations are also provided by volunteers; research and marketing committees, tasting panels for quality control, and functions such the Napa auction are run primarily on a volunteer basis. Winegrower support for initiatives such as open cellars and participation in promotion campaigns is crucial and helps make the territory.

### The Physical and Cultural Territory

Wine tastes best in the territory that produced it. It tastes even better on the estate that grew the grapes and made the wine, especially when complemented with local dishes. The invitation to the territory and into the estates is the most tempting offer winegrowers can make. Consumers not only experience variety, but appreciate it in its physical, cultural, and production context, see the ground it comes from, meet the winemaker, pet their airedale. Mountains and churches are for exploring, chateaux to gawk at, rebels to meet and lots of different wines to taste. Territories offer a diversity and authenticity that corporations can't match. To make the invitation attractive, however, the individual estates and those from the territory as a whole have to be taken care of.

Whether in cars, buses, on bicycles or on a marathon from cellar to cellar, tourists expect a landscape of beautiful vineyards in a harmonious relationship with the natural environment. They don't expect concrete. Reducing the environmental impacts of wineries and vineyards and preventing damage from urban sprawl, road expansion, quarrying and so on are increasingly key issues for winegrower associations. Organic and sustainable winegrower associations often preceded such awareness, and where they once faced resistance, they are now the vanguard of broader collective efforts. All three territories need to preserve important artefacts such as ghost wineries, churches, or imposing chateaux. More importantly they have living traditions like festivals to carry forward, rejuvenate and replace. Such preservation and innovation usually requires collaboration with local, national and international organizations. UNESCO is a particularly popular partner among territories. Coalitions bring together winegrower associations interested in tourist experience and territorial image in distant markets with governments and NGOs interested in the local and global significance of environmental and cultural preservation.

Territory building was not stressed until recently. Winegrowers focused on quantity and quality. Now tourism and building the cultural and environmental image of the territory are key strategies of associations. The importance of this type of territorial building is taking the associations to a new level of sophistication. It is also taking winegrowers to a different form of cooperation. Positive externalities that can be shared by all must be created and maintained. Investments must be made in buildings, maintenance, pollution controls, tasting rooms and other facilities, while restraints are made in vineyard expansion, winery buildings, or (ironically) tourist facilities.

Cooperation for territory building may be elicited with existing funding and regulatory mechanisms; others have and will require new discussions, regulations, and even laws. Fortunately, the primary strength of wine territories is their diversity, and it is likely that the initiatives, leadership and proposals for cooperation will come from the community. The women of Chianti Classico transformed the culture of their territory. Organic and sustainable winegrowers, since initially suffering chemical and critical drift from their neighbours, have diffused information and shown how costs to individual winegrowers and community can be reduced. For those who accept the winegrowers' invitation, perhaps gaining an appreciation of their cooperation and joy in community is the most outstanding experience the visitor will take home.

Whether seeking a living, or chasing a dream, winegrowers need to differentiate. Lack of differentiation results in commodification and the necessity to get out or get bigger. Counter-intuitively, the differentiation of winegrowers has led them to create a new type of organization that will supply them with many economies of scale and scope that they cannot provide by themselves. It has also required them to govern their organization with innovation, flexibility and thoroughness. But it is through that organization that they can extend an invitation to variety. It is an expression of variety that differs in quantity and characteristics from what is offered by large corporations. The winegrowers and their territorial institutions have no perfect system to manage the distribution change or protect their landscape. At the far end of the distribution chain, there are lots of obstacles within food culture and social trends, but also many movements demanding, and a broader to acceptance, for greater variety in choices. The invitation extended by winegrowers is one for both consumers and economic theorists to reflect upon.

# Endnotes

## Notes to Chapter 1

1. Kelvin Lancaster, *Variety, Equity and Efficiency* (New York: Columbia University Press, 1979): 7.
2. Ibid.
3. Ibid.
4. Sébastien Lecocq and Michael Visser, "What Determines Wine Prices: Objective vs. Sensory Characteristics," *Journal of Wine Economics* 1 (2006): 42–56; Eddie Oczkowski, "A Hedonic Price Function for Australian Premium Wine" *Australian Journal of Agricultural Economics* 38 (1994): 93–110; Gunter Schamel, "Geography versus Brands in a Global Wine Market," *Agribusiness* 22 (2006): 363–74.
5. Jean Philippe Perrouty, François d'Hauteville and Larry Lockshin, "The Influence of Wine Attributes on Region of Origin Equity: An Analysis of the Moderating Effect of Consumer's Perceived Expertise," *Agribusiness* 22 (2006): 323–41; G. Malorgio, A. Hertzberg and C. Grazia, "Italian Wine Consumer Behaviour and Wineries Responsive Capacity," paper presented at the 12th EAAE Congress. "People, Food and Environments: Global Trends and European Strategies," Gent (Belgium, August 26–29, 2008) http://ideas.repec.org/p/ags/eaae08/44419.html.
6. Seminal approaches to hedonic and discrete choice analysis, as cited by several wine researchers are: Sherwin Rosen, "Hedonic Prices and Implicit Markets," *Journal of Political Economy* 82 (1974): 34–55; Daniel L. Mcfadden, "The Choice Theory Approach to Market Research," *Marketing Science* 5 (1986): 275–97.
7. Sheena S. Iyengar and Mark R. Lepper, "When Choice is Demotivating: Can One Desire Too Much of a Good Thing?," *Journal of Personality and Social Psychology* 79 (2000) 995–1006; Naresh K. Malhotra, "Information Load and Consumer Decision Making," *Journal of Consumer Research* 8 (1982): 419–30; Itamar Simonson, "The Effect of Product Assortment on Buyer Preferences," *Journal of Retailing* 75 (1999): 347–70.
8. Alexander Chernev, "When More Is Less and Less Is More: The Role of Ideal Point Availability and Assortment in Consumer Choice," *Journal of Consumer Research* 30 (2003): 170–83.
9. Jonah Berger, Michaela Draganska and Itamar Simonson, "The Influence of Product Variety on Brand Perception and Choice," *Stanford Graduate School of Business Research Paper Series* No. 1938 (2006).
10. Cassie Mogilner, Tamar Rudnick and Sheena S. Iyengar, "The Mere Categorization Effect: How the Presence of Categories Increases Choosers' Perceptions of Assortment Variety and Outcome Satisfaction," *Journal of Consumer Research* 35 (2008): 202–15.

11. Dennis W. Carlton and James D. Dana, Jr., "Product Variety and Demand Uncertainty," NBER Working Paper No. 10594 (2004).

12. George A. Akerlof, "The Market for 'Lemons': Qualitative Uncertainty and the Market Mechanism," *Quarterly Journal of Economics* 84 (1970): 488–500; an explanation of how a lack of discrete information reduces consumer evaluation of products. In Akerlof's famous example, the inability of consumers to differentiate between cars of different value, led consumers to evaluate all as "lemons."

13. Philip Nelson, "Information and Consumer Behavior," *Journal of Political Economy* 78 (1970): 311–29.

14. Fiona Scott Morton and Joel Podolny, "Love or Money? The Effects of Owner Motivation in the California Wine Industry," *Journal of Industrial Economics* L (2002): 431–56.

15. Seminal works in this area include: Richard Le Heron, *Globalized Agriculture: Political Choice* (Oxford: Pergamon, 1993); David Goodman and Michael Watts, *Globalizing Food: Agrarian Questions and Global Restructuring* (London: Routledge, 1997); Helena Paul and Ricarda Steinbrecher, *Hungry Corporations* (London: Zed Books, 2003).

16. Pier Paolo Saviotti, "Product Differentiation," in *Markets and Organization*, eds Richard Arena and Christian Longhi (Berlin: Springer, 1998): 487–503; Kelvin J. Lancaster, "The Economics of Product Variety: A Survey," *Marketing Science* 9 (1990), 189–206.

17. Lancaster op. cit., 1979.

18. Richard Schmalensee, "Entry Deterrence in the Ready-to-Eat Breakfast Cereal Industry," *Bell Journal of Economics* 9 (1978): 305–27; Frederic M. Scherer, "The Welfare Economics of Product Variety: An Application to the Ready-to-Eat Cereals Industry," *The Journal of Industrial Economics* 28 (1979): 113–34; John M. Connor, "Breakfast Cereals: The Extreme Food Industry," *Agribusiness* 15 (1999): 247–59.

19. Tim Unwin, *Wine and the Vine: An Historical Geography of Viticulture and the Wine Trade* (New York: Routledge, 1991); Hugh Johnson, *Story of Wine* (London: Mitchell Beazley, 1998); Rod Phillips, *A Short History of Wine* (London: Allen Lane, 2000).

20. Anand Swaminathan, "Resource Partitioning and the Evolution of Specialist Organizations: The Role of Location and Identity in the US Wine Industry," *Academy of Management Journal* 44 (2001): 1169–86.

21. Louis Loubère, *The Wine Revolution in France* (Princeton: Princeton University Press, 1990).

22. Onivins, *Les Vins Français face à la concurrence internationale* (Paris: Onivins, 2003).

23. Philip Kotler and Kevin Lane Keller, *A Framework for Marketing Management* (Upper Saddle River: Pearson International, 2007): 136.

24. Tina Caputo, "Many Wine Consumers 'Overwhelmed': Scan Data Further Defines 'Project Genome' Consumer Segments," *Wines and Vines* (March 10, 2008), http://www.winesandvines.com/template.cfm?section=news&content=53745.

25. Neil Wrigley, "The Consolidation Wave in US Food Retailing: A European Perspective," *Agribusiness* 17 (2001): 489.

26. Annemiek Geene et al., *The World Wine Business, Market Study 1999* (Utrecht: Rabobank, 1999).

27. Barbara Insel, "Distributors and the Information Gap," *MKF Research*, Presentation to Family Winemakers of California (January 22, 2007), http://www.familywinemakers. org/doc_files/InselDist.pdf.

28. Ibid. Rabobank op. cit.

29. Alfredo Manuel Coelho, and Jean-Louis Rastoin, "Globalization du marché du vin et restructuration des enterprises multinationales (Globalization of the wine industry and the restructuring of multinational enterprises)," (paper presented at Oenometrie XI Université de Borugogne, Dijon, May 21–22, 2004).

30. Unwin op. cit.

31. Johnson op. cit.

32. Gunter Schamel, "A Dynamic Analysis of Regional and Producer Reputation for California Wine" (paper presented at Oenometrics X Budapest, Hungary May 22–24, 2003); Gunter Schamel and Kym Anderson K., "Wine Quality and Varietal, Regional and Winery Reputations: Hedonic Prices for Australia and New Zealand," *The Economic Record* 79 (2003): 357–69.

33. S. Landon and C.E. Smith, "The Use of Quality and Reputation Indicators by Consumers: The Case of Bordeaux Wine," *Journal of Consumer Policy* 20 (1997): 289–323; S. Landon and C.E. Smith, "Quality Expectations, and Price," *Southern Economic Journal* 64 (1998): 628–47.

34. Ulrich Orth, "Creating and Managing Regional Umbrella Brands: A Comprehensive Quantative Approach," (paper presented at 2nd International Wine Marketing Symposium, Sonoma State University July 8–9 2005); Tom Shelton, "Product Differentiation," in *Successful Wine Marketing*, eds Kirby Moulton and James Lapsley (Gaithersburg: Aspen, 2001): 99–106.

35. Eric Giraud-Héraud et al., "La regulation interprofessionelle das le secteur vitivinicole est-elle fondée economiquement?," *Bulletin de L'O.I.V.* 813–14 (1998): 1060–84; see also Filippo Arfini, "The Value of Typical Products: The Case of Prosciutto Di Parma and Parmigiano Reggiano Cheese," in *The Socio-Economics of Origin Labelled Products in Agri-Food Supply Chains: Spatial, Institutional, and Co-ordination Aspects*, eds Bertil Sylvander, Dominique Barjolle and Filippo Arfini (Versaillesa: INRA Editions, no. 17-1, 2000): 77–97.

36. Akerlof op. cit.

37. Warren Moran, "The Wine Appellation as Territory in France and California," *Annals of the Association of American Geographers* 83 (1993): 694–717.

38. Sergio Escudero, *International Protection of Geographical Indications and Developing Countries* (South Centre: Trade Related Agenda Development and Equity, working paper 10, 2001).

39. Aaron Kingsbury and Roger Hayter, "Business Associations and Local Development: The Okanagan Wine Industry's Response to NAFTA," *Geoforum* 37 (2006): 596–609.

40. Daniel Gade, "Tradition, Territory, and *Terroir* in French Viniculture: Cassis, France, and Appellation Contrôlée," *Annals of the Association of American Geographers* 94 (2004): 848–67.

41. Morton and Podolny op. cit.

42. Elisa Giuliani, "The Selective Nature of Knowledge Networks in Clusters: Evidence from the Wine Industry," *Journal of Economic Geography* 7 (2007): 139–68.

43. Jean Tirole, "A Theory of Collective Reputations (with applications to the persistence of corruption and to firm quality)," *The Review of Economic Studies* 63 (1996): 1–22.

44. Elinor Ostrom, *Governing the Commons* (Cambridge: Cambridge University Press, 1990); Elinor Ostrom, "Property-rights Regimes and Common Goods: A Complex Link," in *Common Goods*, ed. Adrienne Héritier (Lanham: Rowman & Littlefield Publishers, 2002): 15–28; Elinor Ostrom et al., "Revisiting the Commons: Local Lessons, Global Challenges," *Science* 284 (1999): 278–82.

45. Bertil Sylvander, Dominique Barjolle, and Filippo Arfini, *The Socio-Economics of Origin Labelled Products in Agri-Food Supply Chains: Spatial, Institutional, and Co-ordination Aspects*. Versaillesa: INRA Editions, numbers. 17-1 and 17-2, 2000); Arfini op. cit; Giraud-Héraud et al. op. cit.

46. Dennis Carlton and Jeffrey Perloff, *Industrial Organization* (Boston: Pearson, 2005).

47. Giraud-Héraud, E., Soler, L.G, Tanguy, H., "La relation vignoble-négoce: efficacité de structures verticals différenciées? (The vineyard-merchant relation: efficiency of differentiated vertical structures?)," in *The Socio-Economics of Origin Labelled Products in Agri-Food Supply Chains: Spatial, Institutional, and Co-ordination Aspects*, eds Bertil Sylvander, Dominique Barjolle and Filippo Arfini (Versaillesa: INRA Editions, no. 17-2, 2000): 163–75.

48. John Crespi and Stéphan Marette, "Generica Advertising and Product Differentiation," *American Journal of Agricultural Economics* 84 (2002): 691–701.

49. Giovanni Belletti et al., "Individual and Collective Levels in Multifunctional Agri-culture," http://www.gis-syal.agropolis.fr/Syal2002/FR/Atelier%205/BELLETTI%20 MARESCOTTI.pdf.

50. Bernd Frick, "Does Ownership Matter? Empirical Evidence from the German Wine Industry," *KYKLOS* 57 (2004): 357–86.

51. Jerry Patchell, "Kaleidoscope Economies: Cooperation, Competition and Control in Regional Development," *Annals of the Association of American Geographers* Vol. 86 (1996): 481–506.

52. Alice Feiring, *The Battle for Wine and Love or How I Saved the World from Parkerization* (New York: Harcourt, 2008); Alain Marty, *Ils vont tuer le vin français* (Paris: Ramsay, 2004); Andrew Jefford, *The New France* (London: Mitchell Beazley, 2002); Lawrence Osbourne, *The Accidental Connoisseur: An Irreverent Journey Through the Wine World*, (New York: North Point Press, 2004); Jonathan Nossiter, *Mondovino* (ThinkFilm, 2004).

53. Sarah Whatmore and Lorraine Thorne "Nourishing Networks," in *Globalizing Food*, eds David Goodman and Michael Watts (London: Routledge, 1997): 287–304; Terry Marsden "Food Matters and the Matter of Food: Towards a New Food Governance," *European Society for Rural Sociology* 40 (2002): 20–29; D.C.H. Watts, B. Ilbery, and D. Maye, "Making Reconnections in Agro-food Geography: Alternative Systems of Food Provision," *Progress in Human Geography* 29 (2004): 22–40; K. Morgan, T. Marsden, and J. Murdoch, *Worlds of Food. Place, Power, and Provenance in the*

*Food Chain* (Oxford: Oxford University Press, 2006); D. Maye, L. Holloway, and M. Kneafsey, *Alternative Food Geographies. Representation and Practice* (Amsterdam. Elsevier, 2007).

54. Henk Renting, Terry Marsden and Jo Banks, "Understanding Alternative Food Networks: Exploring the Role of Short Food Supply Chains in Rural Development," *Environment and Planning A* 35 (2003): 393–411.

55. Sylvander, Barjolle and Arfini op. cit.; Gade op. cit.; Elizabeth Barham, "Translating *Terroir*: The Global Challenge of French AOC Labelling," *Journal of Regional Studies* 19 (2003): 127–38; Nick Lewis et al., "Territoriality, Enterprise and Réglementation in Industry Governance," *Progress in Human Geography* 26 (2002): 433–62; Warren Moran, "The Wine Appellation as Territory in France and California," *Annals of the Association of American Geographers* 83 (1993): 694–717; Warren Moran, "Rural Space as Intellectual Property," *Political Geography* 12 (1993): 263–77.

56. Henk Renting, Terry Marsden and Jo Banks op. cit.

## Notes to Chapter 2

1. Reinhard Bendix, *Work and Authority in Industry* (Berkeley: University of California Press, 1974).

2. Nick Lewis et al., "Territoriality, Enterprise and Réglementation in Industry Governance," *Progress in Human Geography* (2002): 433–62. See also, Warren Moran et al., "Economic Organization and Territoriality Within the Wine Industry of Quality: A Comparison Between France and New Zealand," *The Socio-Economics of Origin Labelled Products in Agri-Food Supply Chains: Spatial, Institutional and Co-ordination Aspect*, eds Bertil Sylvander, Dominique Barjolle and Filippo Arfini (Versaille: INRA-Editions, no. 17-1, 2000): 315–28.

3. The basic conditions for a common pool resource are defined more generally and formally as i) "exclusion of beneficiaries through physical or institutional means is especially costly" and ii) "exploitation by one user reduces resource availability to other users" Elinor Ostrom, "Revisiting the Commons: Local Lessons, Global Challenges," *Science* 284 (1999): 278–82.

4. Michael Storper, *The Regional World* (New York: Guilford, 1997).

5. Adapted from Elinor Ostrom, "Property-Rights Regimes and Common Goods: A Complex Link," in *Common Goods*, ed. Adrienne Héritier (Lanham: Rowman & Littlefield Publishers, 2002): 15–28.

6. Elinor Ostrom, *Governing the Commons* (Cambridge: Cambridge University Press, 1990).

7. Fiona Scott Morton and Joel Podolny, "Love or Money? The Effects of Owner Motivation in the California Wine Industry," *Journal of Industrial Economics* L (2002): 431–56.

8. Elizabeth Barham, "Translating *Terroir*: The Global Challenge of French AOC Labelling," *Journal of Regional Studies* 19 (2003): 127–38; Greig Tor Guthey, "Agro-

Industrial Conventions: Some Evidence From Northern California's Wine Industry," *Geographical Journal* 174 (2008): 138–48.

9. Daniel Gade, "Tradition, Territory, and *Terroir* in French Viniculture: Cassis, France, and Appellation Contrôlée," *Annals of the Association of American Geographers* 94 (2004): 848–67.

10. Terry Marsden, "Food Matters and the Matter of Food: Toward a New Food Governance," *Sociologia Ruralis* 40 (2000): 20–29.

11. Gianluca Brunori and Adanella Rossi, "Synergy and Coherence Through Collective Action: Some Insights from Wine Routes in Tuscany," *Sociologia Ruralis* 40 (2000): 409–23.

12. Nick Lewis, "Constructing Economic Objects of Governance: The New Zealand Wine Industry," in *Agri-food Commodity Chains and Globalising Networks*, eds Christina Stringer and Richard LeHeron (Aldershot: Ashgate, 2008): 319–23.

13. Dwijen Rangnekar, *The Socio-Economics of Geographical Indications: A Review of Empirical Evidence from Europe* (Geneva: International Centre for Trade and Sustainable Development, 2004); André Torre, "Collective Action, Governance Structure and Organizational Trust in Localized Systems of Production. The Case of AOC Organization of Small Producer," *Entrepreneurship & Regional Development* 18 (2006): 55–72.

14. Philippe Perrier-Cornet and Bertil Sylvander, "Firmes, coordinations et territorialité. Une lecture économique de la diversité des filières d'appellation d'origine," *Économie Rurale* 258 (2000): 79–89.

15. Bertil Sylvander, Dominique Barjolle and Filippo Arfini, *The Socio-Economics of Origin Labelled Products in Agri-Food Supply Chains: Spatial, Institutional and Co-ordination Aspect* (Versaille: INRA-Editions, no. 17-1, 2000): 315–28.

16. Aaron Kingsbury and Roger Hayter, "Business Associations and Local Development: The Okanagan Wine Industry's Response to NAFTA," *Geoforum* 37 (2006): 596–609.

17. Elisa Giuliani, "The Selective Nature of Knowledge Networks in Clusters: Evidence From the Wine Industry," *Journal of Economic Geography* 7 (2007): 139–68.

18. Jean-Claude Hinnewinkel, *Les terroirs viticoles, origines et devenirs* (Bordeaux, Éditions Féret, 2004).

19. Warrren Moran, "The Wine Appellation as Territory in France and California," *Annals of the Association of American Geographers* 83 (1993): 694–717; Warren Moran, "Rural Space as Intellectual Property," *Political Geography* 12 (1993): 263–77.

20. S. Guillet, "Economics of the Appellation Wine Sector: Observation and Prospects," in *Vine and Wine Economy*, ed. E.P. Botos (Oxford: Elsvier, 1990).

21. For France see Andrew Jefford, *The New France* (London: Mitchell Beazley, 2002); for Italy see Nicholas Belfrage, *Brunello to Zibibbo* (London: Faber and Faber, 2001); for both and others see Tom Stevenson, *The Southeby's Wine Encyclopedia* (London: Dorling Kindersley, 2005).

22. Gérard César, *No. 349 Sénat session extraordinaire de 2001–2002, rapport d'information fait au nom de la commission des affaires éconmiques et du plan (1) par le group de travail (2) sur L'avenir de la viticulture française* (2002), http://www.senat.fr/rap/r01-

349/r01-3490.html; Jean-Pierre Deroudille, *Le Vin Face À La Mondialisation* (Paris: Hachette, 2003); Alain Marty, *Ils vont tuer le vin français* (Paris: Ramsay, 2004).

23. Barnham op. cit.

24. http://www.wineinstitute.org/industry/ava/tenthings/avas_batfblessing.htm.

25. Dominique Barjolle and Bertil Sylvander, *Protected Designations of Origins and Protected Geographical Indications in Europe: Regulation or Policy?* (European Union Contrat Fair CT 95-306: PDO and PGI Products, 2000).

26. Sergio Escudero, *International Protection of Geographical Indications and Developing Countries* (South Centre: Trade Related Agenda Development and Equity, working paper 10, 2001).

27. Trade-related aspects of intellectual property rights, a World Trade Organization agreement.

28. Escudero op. cit.

29. Barjolle, Dominique and Bertil Sylvander, 2000, "Some Factors of Success for Origin Labelled Products in Agri-food Supply Chains in Europe: Market, Internal Resources and Institutions," in *The Socio-Economics of Origin Labelled Products in Agri-Food Supply Chains: Spatial, Institutional, And Co-ordination Aspects*, eds Bertil Sylvander, Dominique Barjolle and Filippo Arfini (Versaillesa: INRA Editions, no. 17-1, 45–71, 2001).

30. http://www.gmabrands.com/news/docs/NewsRelease.cfm?DocID=1184.

31. Anand Swaminathan, "Resource Partitioning and the Evolution of Specialist Organizations: The Role of Location and Identity in the US Wine Industry," *Academy of Management Journal* 44 (2001): 1169–85.

32. Originally described by Michael E. Porter, *Competitive Advantage* (New York: Free Press, 1985); and Michael E. Porter, *The Competitive Advantage of Nations* (New York: Free Press, 1990); but adopted and developed by many other authors.

33. Gerry Gereffi, "The Organization of Buyer-driven Global Commodity Chains: How the US Retailers Shape Overseas Production Networks," in *Commodity Chains and Global Capitalism*, eds G. Gereffi and M. Korzeniewicz (Westport: Praegar, 1994); G. Gereffi, J. Humphrey and T. Sturgeon, "The Governance of Global Value Chains," *Review of International Political Economy* 12 (2005): 78–104; David Hayward and Nick Lewis, "Regional Dynamics in the Globalising Wine Industry: The Case of Marlborough, New Zealand," *The Geographical Journal* 174 (2008): 124–37.

34. Oliver E. Williamson, *Mechanisms of Governance* (New York: Oxford University Press, 1996).

35. Lewis et al. op. cit.

36. Coe N.M. et al., "Globalizing Regional Development: A Global Production Networks Perspective," *Transactions of the Institute of British Geographers* NS 29 (2004): 468–84.

37. E. Giraud-Héraud, L.G. Soler, and H. Tanguy H., "La relation vignoble-négoce: efficacité de structures verticals différenciées? (The vineyard-merchant relation: efficiency of differentiated vertical structures?)," in *The Socio-Economics of Origin Labelled Products in Agri-Food Supply Chains: Spatial, Institutional, and Co-ordination*

*Aspects*, eds Bertil Sylvander, Dominique Barjolle and Filippo Arfini (Versaillesa: INRA Editions, no.1, 2001).

38. Oliver E. Williamson, *The Economic Institutions of Capitalism: Firms, Markets, Relational Contracting* (New York: Free Press, 1985); Williamson (op. cit., 1996).

39. E. Giraud-Héraud et al., *"La regulation interprofessionelle das le secteur vitivinicole est-elle fondée economiquement?"* Bulletin de L'O.I.V. 813–14 (1998): 1060–84; Giulio Malorgio and Cristina Grazia, "Quantity and Quality Regulation in the Wine Sector: The Chianti Classico Appellation of Origin," *International Journal of Wine Business Research* 19 (2007): 298–310; Sylvander et al. (op. cit., 2000).

40. Ann T. Coughlan, Erin Anderson, Louis Stern and Adel I. El-Ansary, *Marketing Channels* 7th Edition (Upper Saddle River: Pearson International, 2006).

41. Dennis Carlton and Jeffrey Perloff, *Industrial Organization* (Boston: Pearson, 2005).

42. Giraud-Héraud et al. op. cit.

43. Forker, O.D. and R.W. Ward, *Commodity Advertising: The Economics and Measurement of Generic Programs* (New York: Lexington Book, 1993); Richard Kohl and Joseph Uhl, *Marketing of Agricultural Products* (Upper Saddle River: Prentice Hall, 2002).

44. John Crespi and Stéphane Marette, "Generica Advertising and Product Differentiation," *American Journal of Agricultural Economics* 84 (2002): 691–701.

45. Stéphane Marette and Angelo Zago, "Quality and International Trade: What Strategies for EU AOC System," paper presented at International Conference Agricultural policy reform and the WTO: where are we headed? (Capri, June 23–26, 2003).

46. Ostrom op. cit., 1990.

47. Jean Tirole, "A Theory of Collective Reputations (with applications to the persistence of corruption and to firm quality)," *The Review of Economic Studies* 63 (1996): 1–21.

**Notes to Chapter 3**

1. Hugh Johnson, *The Story of Wine* (London: Mitchell Beasley, 2002).

2. Nicholas Faith, *The Wine Masters of Bordeaux* (London: Prion, 1999): 35.

3. Bruno Boidron and Marc-Henry Lemay, *Bordeaux Vins & Négoce* (Bordeaux: Éditions Feret, 2000).

4. Enforced through a system known as police des vins; Johnson op. cit., p. 145.

5. Jean-Claude Hinnewinkel, *Les terroirs viticoles, origines et devenirs* (Bordeaux, Éditions Féret, 2004): 27.

6. Johnson op. cit. p. 260.

7. Faith op. cit.

8. Markham Dewey Jr., *1855: A History of the Bordeaux Classification* (New York: John Wiley and Sons, 1997).

9. René Pijassou, *Le Médoc* (Paris: Éditions Tallandier, 1980).

10. Philipe Roudié, *Vignobles et Vignerons du Bordelais* (1850–1980) (Bordeaux: Presses Universitares de Bordeaux, 1988): 143.

11. Roudié op. cit., p. 142.

12. Roudié op. cit., p. 105.

13. Pijassou op. cit., p. 837.
14. Faith op. cit., p. 140.
15. Hinnewinkel op. cit., p. 77.
16. Hinnewinkel op. cit., p. 83.
17. The Association Syndicale des viticulteurs propriétaires de la Gironde (1895) (later the Union girondine des syndicats agricoles 1906) and the Syndicat girondin de défense contre la fraude; Roudié op. cit., 221.
18. Faith 1999 op. cit., p. 126.
19. Hinnewinkel op. cit., p. 130; Faith op. cit., p. 167; Roudié op. cit., p. 296.
20. Roudié op. cit., p. 352.
21. Roudié op. cit., p. 381.
22. Faith op. cit.
23. Gérard César, *No. 349 Sénat session extraordinaire de 2001–2002, rapport d'information fait au nom de la commission des affaires éconmiques et du plan (1) par le group de travail (2) sur L'avenir de la viticulture française* (2002), http://www.senat.fr/rap/r01-349/r01-3490.html; Gérard César, *Vin, Consommation, Distribution: Nouveaux Enjeux, Nouvelles Opportunités?* (Paris: Sénat, 2005).
24. Faith op. cit., p. 249.
25. Roudié op. cit., p. 371.
26. Hinnewinkel op. cit.
27. Stephen Brook, *Bordeaux: People, Power and Politics* (London: Mitchell-Beazley, 2001).
28. Florence Varaine, "Rétrospective d'une semaine marathon," *Union Girondine*, 1000 (2004): 16–21.
29. CIVB (Conseil Interprofessionel du Vin de Bordeaux) *Mémento Économique du Vin de Bordeaux* (Bordeaux: CIVB, 2002a).
30. Lisa Shara Hall, "How Consumers Choose Wine: What Wine Drinkers Recognize About Country and Region of Origin" (June 16, 2009), http://www.winebusiness.com/news/?go=getArticle&dataid=65307.
31. S. Landon and C.E. Smith, "The Use of Quality and Reputation Indicators by Consumers: The Case of Bordeaux Wine," *Journal of Consumer Policy* 20 (1997): 289–323; S. Landon and C.E. Smith, "Quality Expectations, and Price," *Southern Economic Journal* 64 (1998): 628–47.
32. Jean Tirole, "A Theory of Collective Reputations," *The Review of Economic Studies* 63 (1996): 1–21.
33. Eric Giraud-Héraud et al., *"La regulation interprofessionelle das le secteur vitivinicole est-elle fondée economiquement?,"* Bulletin de L'O.I.V. 813–14 (1998): 1060–84.
34. Jancis Robinson, *The Oxford Companion to Wine* 2nd edition (Oxford: Oxford University Press, 1999): 92.
35. Pierre Laborde, *Bordeaux: Métropole régionale, Ville internationale* (Paris: La documentation Française, 1998).
36. Jean-Paul Charrié and Jean Dumas, "Le système économique bordelais Tertiarisation accrue et restructurations industrielles," in *Bordeaux: Métropole régionale, Ville internationale*, ed. Pierre Laborde (Paris: La documentation Française, 1998): 143–168.

37. Monique Perronnet-Menault, "700,000 Bordelais," in *Bordeaux: Métropole régionale, Ville internationale*, ed. Pierre Laborde (Paris: La documentation Française, 1998): 77–106; Bordeaux en chiffres, http://www.bordeaux.fr/ebx/portals/ebx.portal?_nfpb=true&_pageLabel=pgPresStand8&classofcontent=presentationStandard&id=287; Bordeaux, http://en.wikipedia.org/wiki/Bordeaux.

38. Charrié and Dumas op. cit., p. 147.

39. CIVB "Conserver l'intégrité du vignoble de Bordeaux" *Les Fiches Technique; Collection Patrimonie Viticole* 1 (2000).

40. "Approche Comparee des Perspectives et Formes D'intervention des Acteurs Locaux Aupres De L'union Europeenne: Interets Vitivinicoles Du Médoc et Objectif 1 Des Fonds Structurels Dans le Hainaut," *Appel à propositions du GIS-GRALE Axe II : Inter-territorialité et performance institutionnelle* (2001), http://grale.univ-paris1.fr/pgscient/2%20Saez/RappCosta.pdf.

41. Tyler Coleman, *Wine Politics* (Berkeley: University of California Press, 2008): 143.

42. CIVB "La concertation: un préalable indispensable à l'aménagement Collection Patrimoine Viticole N°5" (Julliet, 2001).

43. CIVB 2000 op. cit.

44. CIVB 2002a op. cit.; Contrat D'agglomeration Bordeaux Metropole 2000–2006, http://www.bordeaux-metropole.com/projets/contradagglo.pdf.

45. CIVB "L'espace viticole et les nouvelles règles d'urbanisme," *Les Fiches Technique; Collection Patrimonie Viticole* 8 (Decembre, 2002b).

46. CIVB "Des *terroirs*, victims du béton et de la route?," *Les Fiches Technique; Collection Patrimonie Viticole* 11 (Janvier, 2004).

47. CIVB, Le Vin Bordeaux: l'atout d'une region. *Les Fiches Technique; Collection Patrimonie Viticole* 7 (Avril, 2002c).

48. Alain Marty, *Ils vont tuer le vin français* (Paris: Ramsay, 2004).

49. Oliver Styles, "Sarkozy: I Will Relax Wine Advertising Regulations" (Decanter.com, February 27, 2007).

50. Approche Comparee des Perspectives et Formes D'intervention des Acteurs Locaux Aupres de L'union Europeenne: Interets Vitivinicoles Du Médoc et Objectif 1 des Fonds Structurels Dans le Hainaut 2001 *Appel à propositions du GIS-GRALE Axe II : Inter-territorialité et performance institutionnelle*, http://grale.univ-paris1.fr/pgscient/2%20Saez/RappCosta.pdf.

51. "Dialogue ou manifestation" (Sudouest.com, 2007).

52. Ykems, Réforme de l'OCM viti vinicole: Quel modèle d'organisation pour les Appellations d'Origine? (Paris: Comité National des Interprofessions des Vins à Appellation d'Origine, 2006) available at: http://www.ykems.com/YKPageSiteVin3_en.htm.

53. CIVB, "L'environnement viticole dans la commune," *Collection Patrimoine Viticole* 4 (Avril, 2001).

54. CIVB, *The Bilane Carbone study of the Bordeaux Wine Industry: Another Step in the Environmental Program* (2008).

## Notes to Chapter 4

1. Philipe Roudié, *Vignobles et Vignerons du Bordelais* (1850–1980) (Bordeaux: Presses Universtitaires de Bordeaux, 1988): 220.
2. Ibid., p. 86.
3. Elinor Ostrom, *Governing the Commons* (Cambridge: Cambridge UniversityPress, 1990); Elinor Ostrom, "Property-rights Regimes and Common Goods: A Complex Link," in *Common Goods*, ed. Adrienne Héritier (Lanham: Rowman & Littlefield Publishers, 2002): 15–28; Elizabeth Barham, "Translating *Terroir*: The Global Challenge of French AOC Labelling," *Journal of Regional Studies* 19 (2003): 127–38.
4. Roudié op. cit., p. 257.
5. INAO (Institut national des appellations d'origine des vins et eaux-de-vie), *L'appellation d'origine contrôlée: vins et eaux de vie* (Paris: Euro-impressions, 1985): 27.
6. Syndicat Viticole de Saint-Emilion (SVSE), *Saint-Emilion: The Inside Guide*. (Saint-Émilion: *Syndicat* Viticole de Saint-Emilion, no date).
7. Henri Enjalbert, *Les grands vins de Saint-Émilion, Pomerol, Fronsac* (Paris: Éditions Bardi, 1983).
8. Corine Ruffe, "Les Structures agraires et foncières de la commune de St-Emilion en 1847," in *Saint-Émilion: Terroir et Espace de Vie Sociale*, eds Jacqueline Candau, Philippe Roudié and Corine Ruffe (Bordeaux: Maison des Sciences de L'Homme D'Aquitaine, 1991): 11–98.
9. Philippe Roudié,"Les Viticulteurs de Saint-Émilion," in *Saint-Émilion: Terroir et Espace de Vie Sociale*, eds Jacqueline Candau, Philippe Roudié and Corine Ruffe (Bordeaux: Maison des Sciences de L'Homme D'Aquitaine, 1991): 101–29.
10. Philippe Barbour and David Ewens, *Wine Buyers Guide: Saint Emilion* (London: Wine Buyers Guides, 1990).
11. Bernard Ginestet, *Saint-Émilion* (Paris: Jacques Legrand, 1988).
12. Roudié op. cit.
13. Didier Ters and Stéphane Klein, *Saint-Émilion: Source of Passion and Union* (Saint-Émilion: Éditions Confluences, 1998).
14. Philippe Roudié, "Vignobles et vins de Blaye aux XIX et XXeme siecles," *Cahiers du Vitrezay* 10 (1981): 47–85; Bernard Ginestet, *Cotes de Blaye* (Paris: Jacques Legrand, 1990).
15. Alain Contis, "Vin et Vignerons en Blayais au XVIIIe Siecle," in *Vignes, Vins et Vignerons de Saint-Emilion et d'Ailleurs Eds Jacquelin Candau*, eds Philippe Roudié and Corine Ruffe (Bordeaux: Actes du LI° Congrès d'études régionales de la F.H.S.O., 1990): 127–45.
16. Roudié op. cit., 1981.
17. Bernard Ginestet, *Cotes de Blaye* (Paris: Jacques Legrand, 1990).
18. Roudié op. cit., 1981.
19. Roudié op cit., 1988.
20. Roudié op cit., 1981.
21. Roudié op cit., 1991.

22. Innovative winemakers using intense viticulture to make small volumes of concentrated wines.
23. Ginestet op. cit., 1990, p. 55.
24. Roudié op. cit., 1981.
25. Roudié op. cit., 1981.
26. Roudié op. cit., 1981.
27. Ginestet op. cit., 1990.
28. Roudié op. cit., 1981.
29. Ginestet op. cit., 1990.
30. Marc-Henry Lemay, *Bordeaux and its Wines* (Bordeaux: Éditions Féret, 1998).
31. Markham Dewey, *1855: A History of the Bordeaux Classification* (New York: Wiley, 1998).
32. INAO op. cit.; Warren Moran, "The Wine Appellation as Territory in France and California," *Annals of the Association of American Geographers* 83 (1993): 694–717.
33. Ibid., Ginestet 1990, p. 89.
34. Panos Kakaviatos, "Small Chateaux will Suffer if St Émilion Classification Disappears" (Decanter.com, April 5, 2007).
35. Done for the wineguide Cocks – Éd. Féret which is a guide of all Bordeaux winegrowers, originally for merchants, and is based on the opinions of wine brokers. The guide classified winegrowers as grand cru, cru bourgeois, cru artisan, and cru paysan.
36. S. Guillet, "Economics of the Appellation Wine Sector: Observation and Prospects," in *Vine and Wine Economy*, ed. E.P. Botos (Amsterdam: Elsevier, 1990): 9–14.
37. Barbour and Ewens op. cit., p. 36.
38. CIVB, "Conserver l'intégrité du vignoble de Bordeaux," *Collection Patrimoine Viticole* 1 (2000).
39. Departement De La Gironde, *Communaute de Communes de la Juridiction de Saint-Emilion Commune de Saint-Emilion, Plan Local D'urbanisme (Revision) le Projet d'Aménagement et de Développement Durable* (2006), http://www.saint-emilion.org/.
40. Henri Enjalbert, *Great Bordeaux Wines of St Emilion, Pomerol, and Fronsac* (Paris: Editions Bardi, 1985): 275.
41. Dennis Carlton and Jeffrey Perloff, *Industrial Organization* (Boston: Pearson, 2005).

### Notes to Chapter 5

1. The early 1970s for younger readers.
2. Charles Sullivan, *Napa Wine: A History from Mission Days to Present* (San Francisco: Wine Appreciation Guild, 1994): 135.
3. Ibid., p. 107.
4. William Heintz, *California's Napa Valley: One Hundred Sixty Years of Wine Making* (San Francisco: Scottwall Associates, 1999): 48.
5. Sullivan op. cit., pp. 78–9.
6. Heintz op. cit., p. 139.
7. Sullivan op. cit.

8. Sullivan op. cit., p. 24.

9. Heintz op. cit., p. 80.

10. Heintz op. cit., p. 81.

11. Sullivan op. cit., p. 73.

12. Sullivan op. cit., p. 81.

13. Sullivan op. cit., p. 82.

14. Heintz op. cit., p. 147.

15. Charles Sullivan, *A Companion to California Wine* (Berkeley: University of California Press, 1998): 79.

16. Sullivan op. cit., 1994, p. 144.

17. Sullivan op. cit., 1994, p. 60.

18. Heintz op. cit., p. 84.

19. Sullivan op. cit., 1994, p. 162.

20. Sullivan op. cit., 1994, p. 130.

21. Sullivan op. cit., 1994, p. 114.

22. Sullivan op. cit., 1994, p. 399.

23. Sullivan 1998, p. 48.

24. Heintz op. cit., p. 178.

25. Thomas Pinney, *A History of Wine in America: From Prohibition to the Present* (Berkley: University of California Press, 2005): 356.

26. Sullivan op. cit., 1998, p. 114.

27. Pinney op. cit., 2005, p. 356.

28. Sullivan op. cit., 1994, p. 160.

29. Heintz op. cit., p. 78.

30. Sullivan op. cit., 1994, p. 183.

31. Sullivan op. cit., 1994, p. 1837.

32. Sullivan op. cit., 1994, p. 193.

33. Heintz op. cit., p. 294.

34. Sullivan op. cit., 1994, p. 392.

35. James Lapsley, *Bottled Poetry: Napa Winemaking from Prohibition to the Modern Era* (Berkeley: University of California Press, 1997): 18.

36. Sullivan op. cit., 1994, p. 387.

37. Heintz op. cit., p. 314.

38. Lapsley op. cit.

39. Ibid., p. 7.

40. Sullivan op. cit., 1994, p. 322.

41. James Conway, *The Far Side of Eden* (Boston: Houghton Mifflin Company, 2002): 9.

42. Sullivan op. cit., 1994, p. 365.

43. Mick Winter, *Napa Valley Online*, http://www.napavalleyonline.com/vineyardlist.html; and *Wine Business Monthly* (WineBusiness.com, May 8, 2001).

44. Based on Winter's 70 largest landholders and specifying vineyard owners as those who cultivate their grapes, or who have their vineyard managed for them, or are combined vineyard and winery owners who sell the majority of their grapes. Those that I included as primarily grapegrowers/vineyardists, with their current vineyard acreage,

are: Laird Family Estate, 2000+; Beckstoffer Vineyards, 866; Napa Wine Company, 635; Knoxville Associates, 600; Flora Springs Winery, 580; Jaeger Vineyards, 577; Stage Coach Vineyards, 500; Calplans Partners, 405; UCC Vineyards Group, 400; Juliana Vineyards, 350+; Miller Family, 286; Truchard Vineyards, 270; Usibelli Land Development Corp., 214; Hudson Vineyards, 180; Buchli Station Vineyard, 131.

45. Lapsley 1997, p. 173.

46. James Laube, *California's Great Cabernets* (San Francisco: Wine Spectator Press, 1990): 28.

47. Joseph Phelps, *Joseph Phelps Vineyards: Classic Wines and Rhone Varietals* (Berkeley: Regional Oral History Office, University of California, Berkeley, 1996: 28.

48. Steve Hart, "Labeling of Wine Plan Hit," *The Napa Valley Register* (April 14, 1976).

49. Andrew Beckstoffer, *William Andrew Beckstoffer: Premium California Vineyardist, Entrepreneur, 1960s to 2000s* (Berkeley: Regional Oral History Office, University of California, Berkeley, 1999): 142.

50. Ibid., p. 142.

51. Heintz op. cit., p. 372.

52. Bruce Cass and Jancis Robinson, *Oxford Companion to the Wines of North America*, (Oxford: Oxford University Press, 2000): 70.

53. James Laube, "Napa's Name Down the Drain?," *The Wine Spectator* (May 15, 1999).

54. Andre Tchelistcheff, *Andre Tchelistcheff: Drapes, Wine, and Ecology* (Berkeley: Regional Oral History Office, University of California, Berkeley, 1983): 58.

55. Carneros Quality Alliance, http://www.carneros.org/about.html.

56. James Conaway, *Napa: The Story of an American Eden* (Boston: Mariner, 1990): 294.

57. Conaway op. cit., 2002, p. 295.

58. Julia Siler, *The House of Mondavi: The Rise and Fall of an American Wine Dynasty* (New York: Gotham, 2007).

59. Phelps. *Joseph Phelps Vineyard*, 46.

60. Conaway op. cit., 2002, p. 291.

61. Paul Franson, "Napa Ridge Winery Gains a Home to Match Its Presence," *Wine Business Monthly* (WineBusiness.com, August 2002).

62. "Daniel Sogg and James Laube Beringer Sells Napa Ridge to Bronco Wine Co." (January 3, 2000), http://www.winespectator.com/Wine/Daily/News/0,1145,900,00.html.

63. *Wine Spectator*, "Never Mind Its Court Loss, Bronco Launches a New California" (May 19, 2006), http://www.winespectator.com/Wine/Features/0,1197,3276,00.html.

64. Bill Kisliuk, "Battle over Calistoga Wine Labels has Industry Talking." *Napa Valley Register* (March 17, 2008).

65. A State mandated self-taxing marketing organization directed by the Wine Institute.

66. Lewis Perdue, *The Wrath of Grapes* (New York: Spike, 1999): 32–7. Other issues include: criticism of the Wine Institute handing out corporate welfare; marketing programmes that encourage the consumption of high alcohol wine and high levels of consumption; and for lobbying against and defeating legislation intended to encourage more moderate drinking.

67. Daniel Sogg, "California Legislature Bans Use of Napa Brand Names on Non-Napa Wines," http://www.winespectator.com/Wine/Daily/News/0,1145,1166,00.html.

68. Ibid.
69. Wine Institute, "AVAs can spoil a perfectly good trademark," http://www.wineinstitute. org/ava/tenthings/avas_trademark.htm.
70. Tim Fish, "Sonoma Follows Napa With New Labeling Law" (October 5, 2006), https:// wwws.cigaraficionado.com/Wine/Features/0,1197,3445,00.html.
71. Tchlistcheff op. cit., p. 145.
72. Maynard Amerine, *The University of California and the State's Wine Industry* (Berkeley: University of California, Bancroft Library, 1972): 64.
73. Perdue op. cit., p. 55.
74. In 2003, for example, the American Vineyard Foundation managed to get 808 competing grapegrowers and vintners to contribute $1,178,652.97 to fund cooperative research. The five largest producers in California (which is responsible for about 90 percent of US production) are responsible for two-thirds of the State's total production: Gallo accounts for 30 percent; Canadaigua wine, 16 percent; the Wine Group, 12 percent; Beringer, 7 percent and Mondavi, 5 percent. Yet they accounted for only $325,752.03 or 28 percent of the contributions. Gallo was the biggest donor at 150,135.86, while the Wine Group only handed over $8,000. Other large firms, such as Bronco and Trinchero contributed significant amounts, but the majority of contributions came from medium and small wineries, grapegrowers and their associations. The foundation biennially polls its contributors to determine their research needs and define the use of the funds. It publishes the results of the poll on its website.
75. Reed Fujii, "Anti-grape Pest Fight Continues Scientists Search for Cure, Vaccine for Pierce's Disease," *Stockton Record* (July 13, 2003).
76. Juliane Poirer Locke, *Vineyards in the Watershed: Sustainable Winegrowing in Napa County* (Napa: Napa Sustainable Winegrowing Group, 2002): 109.
77. Perdue op. cit., p. 56.
78. Sullivan op. cit., 1998, p. 233.
79. http://premierwinebrokers.com/index.php.
80. Paul Franson, "Andy Beckstoffer Administers a Dose of Salts to Napa Valley Grapegrowers" (WineBusiness.com, March 18, 2009).
81. Beckstoffer op. cit., p. 151.
82. Paul Franson, "Growers and Vintners Investigate Relations at Vineyard Economics Seminar," *Vineyard and Winery Management* 28 (2002) Sept/Oct (5), 46–9.
83. *Constellation Brands 2008 Annual Report*, p. 3.
84. Barbara Insel, "Distributors and the Information Gap," *MKF Research, Presentation to Family Winemakers of California* (January 22, 2007), http://www.familywinemakers. org/doc_files/InselDist.pdf.
85. Free the Grapes, http://www.freethegrapes.org/research.html#issue.
86. Coalition for Free Trade, http://www.coalitionforfreetrade.org/litigation/index.html#ny.
87. Deborah Steinthal, "Growing Profits by Getting Smaller," *Practical Winery & Vineyard Magazine* (2004) November/December.
88. Sullivan op. cit., 1994, pp. 266–7.
89. Conaway op. cit., 2002, pp. 82–92.

90. ABAG (Association of Bay Area Governments) Regional Housing Needs, Frequently Asked Questions, http://www.abag.ca.gov/planning/housingneeds/faq.htm.

91. ABAG Regional Housing Needs Determination 1999–2006, http://www.abag.ca.gov/cgi-bin/rhnd_allocation.pl.

92. Jillian Jones, "Clash Between State, Local Law Casts Shadow on Ag Preserve," http://www.napavalleyregister.com/articles/2009/05/10/news/local/doc4a0656239dcfa273238345.txt.

93. John Williams, Owner/Winemaker of Frog's Leap Winery cited in "Symposium Sheds Light On Key Ecological Issues Facing Napa Valley Winegrowers," http://www.robertmondavi.com/aboutcompany/release.asp?ReleaseID=35.

94. Conaway op. cit., 2002.

95. Bay Area Alliance for Sustainable Communities, *Faces of Sustainability: Napa County*, http://www.bayareaalliance.org.

96. Motto, Kryla, and Fisher, *Economic Impact of California Wine* (Napa: MKF, 2000), appendix 1.5. A comparison of the San Joaquin and Napa Valley work forces and their direct compensation shows some startling differences. In both vineyard and winery, Napa employs far more people to produce less wine. Not only are salaries of both vineyard and winery workers substantially higher, when average yearly earnings are looked at as a percentage of total sales, San Joaquin workers are not getting a very good deal. The 34 percent difference between a San Joaquin and Napa Valley vineyard worker's share of the bottle's earnings is especially large, but even in the wineries where San Joaquin's larger investments in capital equipment might indicate a capacity to pay more, Napa's winery workers take 12 percent more home. These figures are not simply a comparison of San Joaquin and Napa, however, but point more generally to the impact on employment when producing fine wines instead of bulk wine.

| | San Joaquin | Napa |
|---|---|---|
| Total Sales | 222,080,871 | 195,103,220 |
| Total Acreage | 57,430 | 36,378 |
| Vineyard Payroll | 32,190,616 | 61,347,492 |
| Vineyard Employees | 2,108 | 3,077 |
| Average Yearly Vineyard Earnings (as percentage of sales) | 15,271 (.000068) | 19,937 (.000102) |
| Winery Payroll | 34,181,496 | 192,092,729 |
| Winery Employees | 954 | 5,259 |
| Average Yearly Winery Earnings (as percentage of sales) | 35,829 (.00016) | 36,526 (.00018) |

97. Wine Business Monthly, "Winemakers Compensation Survey," (October 14, 2006), http://www.winebusiness.com/referencelibrary/webarticle.cfm?dataId=45483.

98. Specific example comes from Duckhorn, a medium sized winery with other Northcoast wines.

99. Liz Thach, "Social Sustainability in the Wine Community: Managing for Employee Productivity and Satisfaction," *Wine Business Monthly* Vol. IX, No. 7 (2002).

100. According to MKF's study, the variation within a year is: 1st quarter, 2,961; 2nd quarter, 3,486; 3rd quarter, 3,869; and 4th quarter, 2,528, appendix 1.5.
101. MKF op. cit., 2000, appendix 3.0.
102. Elizabeth Sagehorn, "Pouring it on: tales from the tasting room," *Napa Valley Life* (Nov/Dec, 2002).
103. Heather Osborn, "New Study: Farmworker Housing is Needed – Now: Draft Confirms that Hundreds Use 'Irregular' Housing," *St. Helena Star* (January 31, 2002).
104. Thach op. cit.
105. Robin Lail, "The Wine Auction," in *Napa Valley*, ed. Patton Howell (San Francisco: Saybrook Publishing Co., 2000): 156.
106. "Winemakers Compensation Survey," *Wine Business Monthly* (October 14, 2006) http://www.winebusiness.com/referencelibrary/webarticle.cfm?dataId=45483.

## Notes to Chapter 6

1. Castellina in Chianti, Gaiole in Chianti, Greve in Chianti and Radda in Chianti are entirely within the territory, while parts of Barberino Val d'Elsa, Castelnuovo Berardenga, Poggibonsi, San Casciano Val di Pesa and Tavarnelle Val di Pesa help make up the territory.
2. Maria Concetta Salemi, *Chianti* (Fiesole: Nardini Editore, 1999).
3. Florence by Net, "Chianti, the History," http://www.florence.ala.it/chianti-history.htm.
4. http://www.pbs.org/empires/medici/renaissance/republic.html.
5. http://www.antinori.it/eng/index.php; http://www.frescobaldi.it/en/home.htm?intro=0; Rod Phillips, *A Short History of Wine* (London: HarperCollins, 2000): 101.
6. Giovanni Brachetti Montorselli, *Un Gallo Nero Che Ha Fatto Storia* (San Andrea in Percussina: Consorzio del Marchio Storico Chianti Classico, 1999).
7. The process is the use of dried grapes to rekindle the fermentation process, to reduce bitterness and sweeten the wine.
8. Leo Loubère, *The Red and the White: A History of Wine in France and Italy in the Nineteenth Century* (Albany: State University of New York Press, 1978): 62.
9. Salemi op. cit., p. 52.
10. Loubère op. cit., p. 60.
11. Raymond Flower, *Chianti: The Land, The People, The Wine* (New York: Universe Books, 1979): 198.
12. The self-governance section owes much to Giovanni Brachetti Montorselli, "Un Gallo Nero Che Ha Fatto Storia" (San Andrea in Percussina: Consorzio del Marchio Storico Chianti Classico, 1999).
13. The borders of Chianti were reset in 1967 and an eighth sub-zone added in 1996.
14. Montorselli op. cit., p. 54.
15. Most likely French demands, as France had signed a memorandum of understanding with the Italians in 1948 bringing Italy towards signing the Madrid convention to ensure protection wine appellations.

16. Tom Stevenson, *The New Sotheby's Wine Encyclopedia* (London: Dorling Kindersley, 1997).
17. Giulio Malorgio and Cristina Grazia, "Quantity and Quality Regulation in the Wine Sector: The Chianti Classico Appellation of Origin International," *Journal of Wine Business Research* 19 (2007): 298–310.
18. Thirty years later they would fight the EU acquiescing to the vulgarization of the distinction.
19. Montorselli op. cit.
20. Joseph Bastianich and David Lynch, *Vino Italiano* (New York: Clarkson Potter, 2005): 205.
21. Stevenson op. cit.
22. Salemi op. cit., p. 16.
23. Nicolas Belfrage, *Brunello to Zibibbo* (London: Faber and Faber, 2001): 7.
24. Ibid., p. 13.
25. Silvia Fiorentini, "The Consortium Guarantees Quality," *On-line Monthly of the Consorzio del Marchio Storico Chianti Classico* 1 (2004), www.chianticlassico.com.
26. "Chianti chiaroscuro" (Talkingdrinks.com, June 16, 2005).
27. Malorgio and Grazia op. cit.
28. "Chianti chiaroscuro" op. cit.
29. Nicolas Belfrage, "Tuscan Delights" (Decanter.com, August 1, 2001).
30. Giovanni Montorselli, *Black Rooster Wines* (Tavernale Val di Pesa: Consorzio del Marchio Storico-Chianti Classico, 2001): 7.
31. Carla Binswanger, "Tuscan Aristocracy Outrages Small Vintners with DOC Proposal" (Decanter.com, November 5, 2001).
32. Richard Baudains, "A Cru for Chianti" (Decanter.com, May 1, 2001).
33. Gianluca Brunori and Adanella Rossi, "Differentiating Countryside: Social Representation and Governance Patterns in Rural Areas with High Density: The Case of Chianti Italy," *Journal of Rural Studies* 23 (2007): 183–205; ibid. Montorselli 1999.
34. Ibid.
35. Ibid.
36. Sylvie Haniez, Podere Terreno, interview.

## Notes to Chapter 7

1. Stafford Beer, *Designing Freedom* (Chichester: Wiley, 1994).
2. Benjamin Scheibehenne, *The Effect of Having Too Much Choice* (Dissertation Mathematisch-Naturwissenschaftlichen Fakultät II der Humboldt-Universität zu Berlin, 2008).
3. Benjamin Scheibehenne, Rainer Greifeneder and Peter M. Todd, "What Moderates the Too-Much-Choice Effect?," *Psychology & Marketing* 26 (2009): 229–53.
4. Erik Brynjolfsson, Yu "Jeffrey" Hu and Michael D. Smith, "From Niches to Riches: Anatomy of the Long Tail," *MIT Sloan Management Review* 47 (2006): 67–71; Erik

Brynjolfsson, Yu "Jeffrey" Hu and Michael D. Smith, "Consumer Surplus in the Digital Economy: Estimating the Value of Increased Product Variety at Online Booksellers," *Management Science* 49 (2003): 1580–96; Erik Brynjolfsson, Yu "Jeffrey" Hu and Duncan Simester, "Goodbye Pareto Principle, Hello Long Tail: The Effect of Search Costs on the Concentration of Product Sales," *SSRN* (2007), http://papers.ssrn.com/sol3/papers.cfm?abstract_id=953587.

5.  Andrea Bonaccorsi and Christina Rossi, "Contributing to the Common Pool Resources in Open Source Software: A Comparison between Individuals and Firms," paper presented to *Dynamics of Industry and Innovation Conference* (Copenhagen, June 27–29, 2006), http://papers.ssrn.com/sol3/papers.cfm?abstract_id=430920; Siobhán O'Mahony, "Guarding the Commons: How do Community Managed Software Projects Protect their Work," *Research Policy* 32 (2003): 1179–98.

6.  Mark Harvey, Andrew McMeekin and Alan Warde, *Qualities of Food* (Manchester: Manchester University Press, 2004).

7.  Elizabeth Barham, "Translating *Terroir*: The Global Challenge of French AOC Labelling," *Journal of Regional Studies* 19 (2003): 127–38.

8.  Robert Feagan, "The Place of Food: Mapping Out the 'Local' in Local Food Systems," *Progress in Human Geography* 31 (2007): 23–42.

9.  David Goodman, "Rural Europe Redux? Reflections on Alternative Agro-Food Networks and Paradigm Change," *Sociologia Ruralis* 44 (2004): 3–16; Roberta Sonnino and Terry Marsden, "Beyond the Divide: Rethinking Relationships Between Alternative and Conventional Food Networks in Europe," *Journal of Economic Geography* 6 (2006): 181–99.

10. Elinor Ostrom, *Governing the Commons* (Cambridge: Cambridge University Press, 1990).

11. Tom Hyland, "Soave Producers Finally Reject Top Designation" (Decanter.com, October 6, 2003).

12. Claude Ménard, "On Clusters, Hybrids and other Strange Forms: The Case of the French Poultry Industry," *Journal of Institutional and Theoretical Economics* 152 (1996): 154–83; Randall E. Westgren, "Delivering Food Safety, Food Quality, and Sustainable Production Practices: The Label Rouge Poultry System in France," *American Journal of Agricultural Economics* Vol. 81 (1999): 1107–11.

13. Warrren Moran, "The Wine Appellation as Territory in France and California," *Annals of the Association of American Geographers* 83 (1993): 694–717; Warren Moran, "Rural Space as Intellectual Property," *Political Geography* 12 (1993): 263–77; Barham op. cit.

14. Gernot Grabher, "The Weakness of Strong Ties: The Lock-in of Regional Development in the Ruhr Area," in *The Embedded Firm*, ed. Gernot Grabher (London: Routledge, 1993).

15. William Echikson, *Noble Rot* (New York: W.W. Norton & Company, 2004): 200.

16. Giulio Malorgio and Cristina Grazia, "Quantity and Quality Regulation in the Wine Sector: The Chianti Classico Appellation of Origin," *International Journal of Wine Business Research* 19 (2007): 298–310.

# Index